PRACTICAL STATISTICS FOR PETROLEUM ENGINEERS

Practical statistics for petroleum engineers

L. T. Stanley

PETROLEUM PUBLISHING COMPANY
Tulsa · 1973

ACKNOWLEDGMENT

Appreciation is expressed to Union Oil Co. of California under whose auspices most of the work was done and to Dr. L. J. O'Brien for his help and suggestions in preparation of the manuscript.

The Petroleum Publishing Company
211 South Cheyenne
Tulsa, Oklahoma 74101, U. S. A.

Library of Congress Card Number: 72–95444
International Standard Book Number: 0–87814–019–0
Printed in U. S. A.

To the Stanley gang:
Barbara, Tom, Tim, Lois and James

Contents

Preface ix
Typical problems where statistical methods may be applied xiii

1 SINGLE VARIABLE SYSTEMS 3

Introduction to Population and Probability 3
What is an Average? 4
Other Measures of Central Tendency 9
Frequency Distributions 10
Measures of Dispersion 12
Normal Distribution 15
Non-Normal Distributions 20
Transformations 22
Coding 24
Weighting 24
Graphical Methods 29
Sampling 34

2 CONFIDENCE LIMITS AND TESTS OF SIGNIFICANCE 39

Introduction to Statistical Tests 39
Confidence Limits 42
Tests of Significance—The t-Test 46
Tests of Significance—The Chi-Square Test 50
Tests of Significance—The Contingency Test 55

3 ANALYSIS OF VARIANCE AND MULTIPLE COMPARISONS OF MEANS 59

Analysis of Variance—General Principles 59
Bartlett's Test for Homogeneity of Variances 63
Analysis of Variance—One-way Classification 64
Analysis of Variance—Two-way Cross-Classification 74
Analysis of Variance—Interaction Effect 77
Analysis of Variance—Cross-Classification with Disproportionate Observations 81
Analysis of Variance—Models I and II 83
Comparison of More Than Two Means 84
Relation of Statistical Tests 86

4 LINEAR RELATIONSHIPS BETWEEN VARIABLES 88

Correlation Coefficient 89
Rapid Correlation Methods 90
Linear Regression 93
Matrix Methods for Regression Analysis 100
Example Problem—Linear Regression 101
Example Problem—Curvilinear Regression 105
Regressions by Graphical Methods 112

5 MULTIPLE CORRELATION AND REGRESSION 116

Introduction 116
Multiple Regression 117
Multiple Correlation 124
Design of Experiments 130

APPENDIX A—STATISTICAL TABLES 139

Table No.
A-1 Percentiles of the t Distribution 141
A-2 Percentiles of the χ^2 Distribution 142
A-3 F Distribution, Upper 5% Points 143
A-4 F Distribution, Upper 1% Points 144
A-5 Percentiles of the Distribution of $q = w/s$ 145
A-6 Fraction of Variance Removed by Correlation (r^2) 149
A-7 Rank Correlation Coefficients 149

APPENDIX B—MATRIX METHODS 151

Nomenclature and General Application of Matrices 151
Least Squares Regression—The Matrix of Normal Equations 155
The Abbreviated Doolittle Method 156

Preface

The petroleum engineer, whether he is engaged in drilling, production, reservoir engineering, or economics, each day is confronted with data. These data are the only facts the engineer has about the environment he is trying to describe, understand, and control; consequently it is essential for him to evaluate his data before he proceeds to draw conclusions and act upon them.

Perhaps petroleum engineering relies upon measurement and observation more than most other types of engineering because of the complexity of the environment with which it deals. Scores of parameters, sometimes independent but more often interrelated, are measured and reported to the engineer each day, and he must treat the results in the context of natural and physical laws in solving his problems.

Since the petroleum engineer is faced with the problem of drawing conclusions under uncertainty, he needs every tool he can get to increase his confidence, or at least to measure his uncertainty. There are many such tools available to him in the field of statistical analysis.

All engineers are engaged in statistical analysis, though they certainly don't think of themselves as statisticians. But each time they take the average of a set of numbers or draw a curve through plotted points they are determining the *central tendency* of the data. The average, itself, is a *statistic,* as a statistic is, by definition, any value computed entirely from a sample.

It is the purpose of this book to describe briefly statistical methods that may be useful to the petroleum engineer and to provide examples taken from actual problems. Whenever a statistical term is introduced, it is italicized and defined. Emphasis is placed on short-cut and graphical methods, because the aim is to present statistical methods as tools for the engineer.

It is not expected that this brief presentation will transform any engineers into instant statisticians, but perhaps it will rescue some from the wearisome habit of emphasizing the insignificant. Statistical methods provide an objective procedure for organizing and evaluating data and, used objectively, they can help the over-worked engineer "put the grease where the squeak is."

The example problems may seem weighted in favor of reservoir engineering. Actually, a survey of the literature has shown that most of the applications that have been made are in reservoir engineering. This is in part due to the large quantities of core data handled by reservoir engineers and in part because of the inherent variability of reservoir data. The statistical methods are the same, though, whether one is handling drilling, production, or cost data as they are for reservoir data.

High-speed digital computers have now made it possible to process huge quantities of data and compute statistical parameters that previously would have been prohibitive in time and manpower. For example, a unitization committee that is faced with the problem of handling thousands of pieces of core data can process all of them by machine in less time than it formerly would have taken to study a small sample of the total.

The example problems given in these articles are solved by hand calculation, but it should be remembered that each can be written as a relatively simple computer program and the quantity of data that can be handled by the machine is almost unlimited.

Most computer centers have a variety of statistical programs in their libraries that they offer as a service to their customers. Thus it is likely that the engineer who has access to a computer will prefer to let the computer do some of the more tedious operations that are outlined here, especially in the area of multiple regression and correlation. It is not the purpose of this book to encourage the engineer to expend valuable time doing hand calculations that could be done more quickly and accurately by machine. Rather, its purpose is to introduce him to some of the things that can be done with engineering data in the field of statistics and give him an adequate foundation in methodology so that he will know what's behind the "black box" solution and will not have to accept it uncritically.

The work is organized into five parts, with each part covering those operations that seem logically to go together. The arrangement has been made so that underlying concepts are explained when they are applied; therefore, the simpler problems have been taken first.

No attempt has been made to delve into statistical theory. Such derivations as are included have been shown only in sufficient detail to provide the reader with some background for understanding a particular method. It is suggested that those who are interested in deeper study of the subject should seek out the general statistical references listed at the end of the book. Also, a bibliography will be found at the end of each chapter listing articles from oil-industry literature describing applications of methods presented in that chapter.

Lyn T. Stanley

Dallas, Texas
April 1973

Typical problems where statistical methods may be applied

Problem	Method to use	Method described on page
A total of 30 permeability measurements are available from a well. What is the best single value to represent the permeability?	Averages and other measures of central tendency.	3–9
How are the 30 permeabilities scattered, and what does this imply about the distribution of permeability within the reservoir?	Histogram; or cumulative relative frequency plotted on normal probability paper.	10–12 and 29–34
Do the permeabilities approach a symmetrical distribution, or are there a disproportionate number of high or low values?	Cumulative relative frequency plotted on normal probability paper.	29–34
Assume the 30 permeabilities are distributed so that they cover a wide range of values and there are a disproportionate number of small or large values. How can the distribution be made symmetrical?	Transformation.	22
Bottom-hole pressures are to be used in statistical calculations involving squares and cross-products. How can these numbers be reduced to more manageable size?	Coding.	24
Reservoir pressures in the early life of a field are not considered as reliable as more recent measurements. How can this be taken into account in statistical computations?	Weighting.	24–29
We have computed the mean and standard deviation of 30 permeabilities. These are estimates of the true mean and the true standard deviation. How much confidence can we place in these estimates?	Confidence limits.	42–46

Problem	Method to use	Method described on page
A group of porosity measurements are available from each of two reservoir strata, and the average of each group has been computed. On the basis of these averages, are the true porosities of the two strata different?	The t-Test.	46–50
The accident records of three districts have been compiled. Do any of the districts have a significantly better or worse record than the average?	Chi-Square Test or Contingency Test.	50–58
A series of bottom-hole pressure measurements have been made with two pressure recorders. Are the two variances significantly different or could one say that both bombs are equally as precise.	The F-Test.	59–60
A large quantity of core data has been taken from a field. Did all the cores come from similar strata, or are there several dissimilar strata in the reservoir?	Analysis of variance.	59–77
A bottom-hole pressure recorder is being calibrated. How can one determine whether the accuracy is affected by temperature and pressure and whether the temperature and pressure effects are independent?	Analysis of variance—Interaction effect.	77–81
Several groups of gas-oil ratio measurements have been made. Are the variances of the groups essentially similar?	Bartlett's Test for homogeneity of variances.	63; 67–69
Several porosity averages have been computed. Which of these probably came from the same stratum and which from different strata?	Multiple comparisons of means.	84–86
Data on oil-zone depths and produced oil gravity are available from a certain area. Is there a relationship between API gravity and depth?	Scatter diagram; or rank correlation test; or correlation coefficient.	89–93
Solution gas-oil ratio is shown by experiment to be linearly related to pressure. Laboratory measurements of solution GOR have been made at several different pressures. What is the equation of the line that fits the data best? How much variance remains after the best line is fitted to the data? What are	Linear Regression.	93–105 112–114

Problem	Method to use	Method described on page
the confidence limits on an estimate of GOR?		
It is desirable to fit the best curve to production-decline data for the purpose of extrapolating the trend for primary reserve estimates. Production rate is not a straight-line function of time. How can the best line be fitted? How much variance remains after the best line has been fitted? What are the confidence limits on an estimate of production rate?	Transformation and Linear Regression.	22; 93–105
Fluid analyses are available from several samples taken from a particular reservoir. It is necessary to get the equation of the average volume factor curve for use in a material-balance computer program.	Curvilinear (polynomial) regression.	105–114
Permeability, porosity, connate-water saturation, and cementation factor have been measured on a number of cores. Are the four properties related? What is the relative importance of each?	Multiple Correlation.	124–130
Assume the four properties stated above have been found to be related. How can one predict connate-water saturation if he knows the permeability, porosity and cementation factor? What will be the confidence limits on the prediction?	Multiple Regression.	117–124
How can one plan an experiment or a series of measurements involving several parameters so as to get the most information for the least expenditure?	Design of Experiments.	130–137

PRACTICAL STATISTICS FOR PETROLEUM ENGINEERS

1

Single-variable systems

Introduction to population and probability

If we discuss the permeability of a reservoir, we are talking about a property that we never can evaluate with certainty, because there are an infinite number of pieces of rock, each with its own permeability, that make up the reservoir.

When we speak of the permeability in such a general sense, we are referring to the permeability population. According to P. O. Johnson's * definition, "The principal function of statistics is to describe certain characteristics of mass phenomena and repetitive events. From the theoretical point of view, unlimited sequences of events or of similar observations are referred to as *statistical universes* or *populations* or *collectives*."

Another way to define a population is that it is the hypothetical set of all possible observations of the type being investigated. Any finite set of observations must be regarded as a *sample* drawn from the population. Thus, a set of core permeability measurements would be a permeability sample.

Samples permit us to draw *inferences* about the population, but in no case can a sample be considered a complete description of a population. However, as the sample size increases, its properties resemble more and more closely those of the parent population, provided that the samples may be considered as *randomly selected* from the popula-

* References are given at the end of the chapter.

tion. In random selection each member of the population has an equal chance of being selected each time a new observation is made.

Statistical methods, then, are concerned with the study of variation, and any value of a given set of elements is considered a *variable*. By this definition permeability would be a variable. A convenient way to describe the variation in a body of data is by a *frequency distribution*. The frequency distribution may be presented in the form of a *histogram,* Fig. 1.1, or a *frequency curve,* Fig. 1.2.

The more commonly-used form of the histogram utilizes proportional frequencies instead of actual frequencies; that is, the occurrence of a given value of the variable is shown as a fraction of the total number of observations comprising the sample. The proportional frequency curve is sometimes called the *probability density distribution,* because the probability of occurrence of a given value of the variable is defined as the proportional frequence with which it occurs *in a large number of observations*.

Engineering applications require that the principal features of a distribution be described numerically. Mainly we are interested in a measure of location—the *arithmetic mean*—and a measure of spread—the *standard deviation*. In the sections following, the concepts of location and spread, or dispersion, will be discussed in some detail, and methods for numerical description of single-variate data will be discussed.

What is an average?

The most familiar statistical tool is the *average*. Averages are often used as simple numerical descriptions of groups of numbers. As all engineers know, it is entirely possible that no member of a group will have a value exactly equal to the average. The average, then simply gives a measure of the *central tendency* (or *"location"*) of the group. The most well-known type of average is the *arithmetic average,* or *mean*.

The arithmetic mean is simply the sum of all the observations divided by the number of observations, thus

$$\overline{X} = \sum_{i=1}^{N} x_i/N \qquad (1.1)$$

where $\overline{X}$ is the mean of a sample containing N observations of x_i. The arithmetic mean is the most widely-used measure of the location of a set of data, but while it is probably the most useful of all statistics, other properties of the data are of nearly equal importance, as will be seen in later sections.

Throughout this text, the bar above a symbol will be used to denote an average or mean; for example, $\overline{\kappa}$ would represent average permeability, $\overline{\phi}$ would be average porosity, etc. Likewise, N will be used to denote the total number of measurements in a sample.

The computation of the arithmetic mean is straightforward, but where N is large, several aids can be employed to reduce the labor and chance for error. These are as follows:

(a) If the observations consist of several significant figures, they can be converted to smaller numbers by subtracting from each a constant quantity and restoring this at the end of the computation. If the constant x_o is subtracted then,

$$\overline{X} = \sum_{i=1}^{N} x_i/N = x_o + \sum_{i=1}^{N} (x_i - x_o)/N \tag{1.2}$$

For example, the arithmetic average of the numbers 95, 92, 98, 94 and 92 could be found by subtracting 90 from each number and averaging the remainders then adding 90, as

$$\overline{X} = [(5 + 2 + 8 + 4 + 2)/5] + 90 = 94.2$$

(b) If the numbers are decimals, they may be multiplied or divided by a power of 10. If they are fractions, they may be expressed as their common denominator, so that

$$\overline{X} = \sum_{i=1}^{N} x_i/N = 1/h\ [\sum_{i=1}^{N} x_i'/N] \tag{1.3}$$

For example, the arithmetic mean of the numbers .00024, .00136, .00057, .00101 and .00098 can be calculated

$$(0.24 + 1.36 + 0.57 + 1.01 + 0.98)/5 \div 1{,}000 = 0.000832$$

and the arithmetic mean of 1/4, 3/4, 3/16, 1/8 and 3/8 can be calculated as

$$[(16 + 48 + 12 + 8 + 24)/5] \times (1/64) = 22/64 = 11/32$$

(c) When N is very large, a long summation can be avoided by grouping the data. There will be some loss of accuracy, but this will be small if the operation is done properly. The procedure is to sort the data in groups of suitable size and treat each group as though all the observations in it had the value of the average of that group. If the number of groups is large enough, the inaccuracies will be negligible. The optimum number of groups that can be used safely and profitably is between 10 and 20.

If a number, k, of samples, of size n_i, is available and each has had its arithmetic mean calculated, the grand mean can be expressed as

$$\overline{X} = (n_1\bar{x}_1 + n_2\bar{x}_2 + \ldots + n_k\bar{x}_k)/N = \sum_{i=1}^{k} n_i\bar{x}_i/N \qquad (1.4)$$

If all samples are the same size, it becomes

$$\overline{X} = 1/k \sum_{i=1}^{k} \bar{x}_i \qquad (1.5)$$

As an example of grouping consider porosity data from 10 wells. If an average porosity is desired it is necessary only to average the averages, with proper weighting for number of samples in each set.

TABLE 1A

Well 1	Well 2	Well 3	Well 4	Well 5	Well 6	Well 7	Well 8	Well 9	Well 10
8.0	18.2	28.4	4.0	17.6	2.0	28.2	17.6	8.2	15.0
12.5	24.3	23.2	6.2	18.3	3.2	30.1	17.3	9.8	16.0
13.2	$\bar{\phi}_2 = 21.2$	18.2	3.8	17.4	6.6	$\bar{\phi}_7 = 29.2$	17.9	10.2	18.2
9.1		17.6	2.2	16.2	5.5		18.1	11.3	$\bar{\phi}_{10} = 16.4$
$\bar{\phi}_1 = 10.7$		25.0	$\bar{\phi}_4 = 4.0$	13.8	8.1	$n_7 = 2$	$\bar{\phi}_8 = 17.7$	13.1	
	$n_2 = 2$	$\bar{\phi}_3 = 22.5$		$\bar{\phi}_5 = 16.7$	10.0			$\bar{\phi}_9 = 10.5$	$n_{10} = 3$
			$n_4 = 4$		$\bar{\phi}_6 = 5.9$		$n_8 = 4$		
								$n_9 = 5$	
$n_1 = 4$		$n_3 = 5$		$n_5 = 5$					
					$n_6 = 6$				

$$\bar{\phi} = [(4)(10.7) + (2)(21.2) + (5)(22.5) + (4)(4.0) + (5)(16.7) + (6)(5.9) + (2)(28.2) + (4)(17.7) + (5)(10.5) + (3)(16.4)]/40 = 14.04$$

The ungrouped mean is 14.09.

In actual practice grouping would not be worthwhile in this problem because there are only 40 observations; but, if there were upwards of

50 core porosity measurements from each well, considerable advantage would be gained from grouping. Also, in this type of problem it is usually desirable to separate the well averages so that areal trends can be studied.

The devices described in (a), (b) and (c) can, of course, be used in combination.

There are several other types of averages any of which may be more meaningful, in some instances, than the arithmetic average. The *weighted average* is familiar to all petroleum engineers. This is an average obtained by application of an appropriate weighting factor to each observation so that

$$\overline{X}_W = [\sum_{i=1}^{N} w_i x_i]/\sum w_i \tag{1.6}$$

where w is the weighting factor. The core-porosity example represents a weighted average.

Other familiar examples of weighting are the contour map, which provides area weighting factors to properties (elevation, thickness, porosity, pressure, etc.) that vary areally, and the technique used by geologists and engineers to weight core porosities by assigning a representative thickness to each sample. The latter is illustrated by the following typical set of porosities from core plugs.

TABLE 1B

Porosity, % ϕ	Thickness, ft w = h	ϕh
25.2	1.0	25.2
21.2	1.5	31.8
19.6	.9	17.6
18.3	.7	12.8
15.2	.6	9.1
13.4	1.3	17.4
12.6	1.2	15.1
11.2	1.8	20.2
16.3	.5	8.2
25.0	.3	7.5
21.8	1.0	21.8
	10.8	186.7

The weighted average porosity, $\overline{\phi} = 186.7/10.8 = 17.3$.

Averages that are perhaps less well known to most engineers are the *harmonic mean* and the *geometric mean*. The harmonic mean is the reciprocal of the arithmetic mean of the reciprocals i.e.,

$$1/\overline{X}_H = N/\sum_{i=1}^{N}(1/X_i) \tag{1.7}$$

The geometric mean is the Nth root of the product of N observations, defined mathematically as

$$\overline{X}_G = (x_1x_2 \ldots x_N)^{1/N} \tag{1.8}$$

or

$$\log \overline{X}_G = 1/N \sum_{i=1}^{N} \log x_i \tag{1.9}$$

The use of the geometric or harmonic mean amounts to a *transformation* of the variable to log x or 1/x, respectively and the calculation of the arithmetic mean of the new variable. More will be said about transformations later but, in general, the harmonic mean is useful wherever the averaged variable enters a computation as a reciprocal; and the geometric mean is useful when the averaged variable covers a very wide range (several orders of magnitude).

The various types of averages have been studied intensively by engineers and researchers faced with the problem of averaging core permeabilities. Their problem is twofold:

(1) Core plug permeabilities vary widely (as is evident when one considers that plug permeabilities ranging from 1 to 1,000 md within a single formation are not unusual) and,

(2) The degree and type of heterogeneity within the reservoir is difficult to express in terms of its effect upon flow models of the reservoir.

Cardwell and Parsons were the first to suggest that limits on reservoir performance might be expressed in terms of different averages of permeability, k.

They pointed out that the arithmetic mean k (weighted for thickness, if necessary) should represent the upper bound of performance and the harmonic mean should represent the lower limit. In a physical sense the arithmetic average is equivalent to the effective permeability of a completely stratified system composed of parallel layers, while the harmonic mean is equivalent to the effective permeability of a system in which samples are arranged in series. Reservoirs composed of samples arranged randomly could be expected to have performance intermediate between these two extremes.

Katz and Tek have tested this hypothesis by use of a mathematical model; Warren, Skiba and Price have tested it by use of pressure buildup data. Katz and Tek concluded that the weighted arithmetic average does represent the upper bound under both steady and unsteady state conditions provided it is assumed there is infinite vertical permeability and therefore, complete crossflow; but that stratification with no crossflow is a more valid representation of the lower bound than is the series configuration represented by the harmonic mean. They also concluded that no single constant mean permeability may be used to represent the performance of a stratified system for all values of time.

The work of Warren et al. showed that the average permeability computed from buildups usually fell somewhere between the arithmetic mean and the harmonic mean, as Cardwell and Parsons anticipated. This work also showed that there was good correlation between the geometric mean k and k from the pressure buildup.

The purpose of all this discussion of averages is to emphasize some of the hazards the engineer must be prepared to face if he uses them indiscriminately. They are the most useful of statistical tools, but the user should always be aware of their limitations. The wide difference between the various averages that can be found in highly-variable data is illustrated by the typical set of core-plug permeabilities in Table 1C.

The various averages obtained from these data are:

Arithmetic mean = 399.9/25 = 16.0 md
Harmonic mean = 25/15.8977 = 1.57 md
Log_{10} geometric mean = 17.238/25 = 0.689
Geometric mean = 4.88 md

Other measures of central tendency

Two other measures of the central tendency that are sometimes useful are the *median* and the *mode*. The median is the value occurring

TABLE 1C

k, md	1/k	$\log_{10}k$
1.4	0.7143	0.146
12.0	0.0833	1.080
0.6	1.6667	−0.222
34.0	0.0294	1.532
41.0	0.0244	1.613
0.5	2.0000	−0.301
61.0	0.0164	1.785
2.2	0.4545	0.343
6.3	0.1587	0.799
8.0	0.1250	0.903
8.0	0.1250	0.903
0.2	5.0000	−0.699
4.7	0.2128	0.672
14.0	0.0714	1.146
2.1	0.4762	0.323
0.4	2.5000	−0.398
7.0	0.1428	0.845
45.0	0.0222	1.654
31.0	0.0322	1.492
2.3	0.4348	0.362
6.2	0.1613	0.792
101.0	0.0099	2.004
9.7	0.1031	0.986
1.0	1.0000	0.000
0.3	0.3333	−0.522
399.9	15.8977	17.238

midway in the data when the data are arranged in order of value. The mode is the value of the variable which occurs most frequently. To measure the median and mode, one must know something about the *frequency distribution* of the variable.

Frequency distributions

Anyone who has done experimental work knows that when he makes N independent measurements of a variable he cannot expect to get N identical values. He also knows intuitively that if each measurement is completely independent of every other measurement, the values will vary randomly from one measurement to the next, and most of the time they will vary within certain limits depending upon the precision of the measuring device and the skill of its operator.

Measurement error usually accounts for only a part of the total variation in a set of measurements, because there are natural variations within the population from which the measurements (or "samples") are taken. For example, if we set out to estimate the porosity of a rock formation, we not only have to contend with measurement error (or reproducibility) but also with the change of porosity from one rock sample to the next.

Sometimes it is enlightening to apply a descriptive device to data so that we can see how they vary. This device, called the *frequency distribution,* was mentioned in the introductory section. As an example, suppose an engineer wants to study the performance of a certain type of rock bit in a given formation. He searches the drilling records to get all data pertinent to his problem and obtains the information given in Table 1D.

TABLE 1D

Bit No.	Footage drilled
1	69
2	72
3	123
4	135
5	95
6	89
7	139
8	109
9	125
10	80
11	104
12	53
13	76
14	102
15	90
16	116
17	105
18	108
19	115
20	110

Results tabulated in order of occurrence—as above—vary erratically, and with a large number of observations it is difficult to derive any conclusions about distribution and spread. The simplest method of arranging the data is to divide the whole range of variation into a

TABLE 1E

Range of footages	Number of observations	Proportion of observations
50–69	2	0.100
70–89	4	0.200
90–109	7	0.350
110–129	5	0.250
130–149	2	0.100
Total	20	1.000

number of groups of equal size, count the number of observations falling in each group and write these in the form of a frequency table. The data arranged in this manner are shown in Table 1E.

Column 2 of Table 1E gives the frequency of occurrence of various footages, and column 3 shows the frequency as a proportion of the total. A plot of proportional frequency versus footage would be the frequency distribution of footage. The frequency distribution is sometimes called the *probability density distribution.*

A graphical device called the *histogram* often is used to approximate the frequency distribution of samples. In this diagram the horizontal axis is divided into segments corresponding to the ranges of the groups. On each segment a rectangle is constructed whose area is proportional to the occurrence of the group. A histogram of the data from Table 1E is shown in Fig. 1.1.

The number of increments to be used in constructing a histogram will depend upon size of sample and the degree of detail needed. Some numerical estimates that can be made from the histogram will be described later under "Graphical Methods."

Measures of dispersion

It has been emphasized that variation is a characteristic of all data. Experience tells us that the degree of variation will, for the most part, determine the measure of confidence we will have in any averages we may derive from a set of data.

There are several measures of dispersion which may be applied to give the investigator a quantitative idea of the variation. These are: *range, standard deviation,* and *coefficient of variation.*

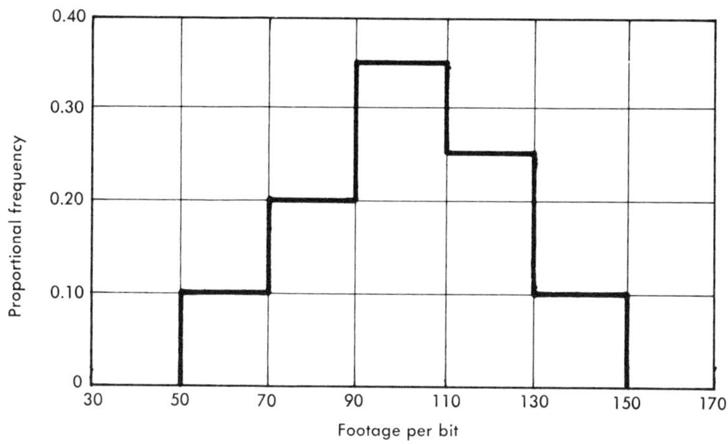

FIG. 1.1 A histogram showing performance of 20 rock bits.

The range is the simplest of all measures of dispersion. It is simply the difference between the highest and lowest values of the variable in a sample. The range of a large sample obviously provides little information, since it says little about the intermediate observations, but for small samples the range will probably convey a large proportion of the information contained in the sample. A method for estimating the standard deviation by using the range of small samples is described in most statistics textbooks.

Standard deviation is well known to most engineers. It is often referred to as the *root mean square*. The standard deviation of a sample is given by

$$s = \sqrt{[\sum_{i=1}^{N}(x_i - \overline{X})^2]/(N - 1)} \qquad (1.10)$$

Equation 1.10 is not the most convenient form for desk computation where a large number of measurements are involved. A more convenient form is

$$s = \{[\Sigma x_i^2 - (\Sigma x_i)^2/N]/(N - 1)\}^{1/2} \qquad (1.11)$$

One of the devices (for simplified computation of the arithmetic mean) described previously may have been used, in which case it would be necessary to compensate for this in calculating the standard deviation. If a constant was subtracted from each number in the set, no correction would be necessary to compensate for this in calculating the standard deviation. But if each number were multiplied by a constant, h, it would be necessary to divide the corresponding standard deviation by h to get the true value. If grouping had been used the equation would become

$$s = \{[\Sigma n_i \overline{X}_i^2 - (\Sigma n_i \overline{X}_i)^2/N]/(N - 1)\}^{1/2} \tag{1.12}$$

The *variance* is another term often encountered in statistical work. The word "variance" is used quite loosely in engineering conversation, but in statistics it has a specific meaning. It is simply the term inside the brackets, or the square of the standard deviation.

The divisor (N − 1) represents the number of *degrees of freedom* of the estimate of variance. If there are N observations, there are (N − 1) independent comparisons, $(x_i - \overline{X})$, that can be made. Only (N − 1) of these are independent because $\overline{X}$ is calculated from the observations and this calculation "uses up" one degree of freedom. If only a few results are available the degrees of freedom are small and the estimate of variance will be of low precision; likewise, the greater the number of observations the higher will be the precision of the estimate.

For the mean, the precision depends on the number of results used in calculating it, but for the variance the precision depends on the number of independent differences in the data from which the variance is calculated. To illustrate the reasoning behind this, let us assume that one measurement is made from a population so that we have a sample composed of one observation. In this single measurement we have an estimate of the population mean (though not a very good one), but what can we say about the variance of the population? Apply the square of Equation 1.11 to the problem, and the result is

$$s^2 = [\Sigma x_i^2 - (\Sigma x_i)^2/N]/(N - 1) = 0/0,$$

which is indeterminate, as logic demands it should be. If we had divided by N, we would have calculated a variance of zero, and we know this is untrue. If a second measurement is made, then the denominator, N − 1, will be equal to 1, because we now have one independent comparison, $x_1 - x_2$, to use as an estimate of population variance. With three observations there would be only two independent comparisons, because when two are given, say $x_2 - x_1$ and $x_2 - x_3$, the third is implied because $x_1 - x_3 = (x_2 - x_3) - (x_2 - x_1)$.

The discussion up to now has concerned the spread of results around a point (the mean) but the same reasoning applies to the spread of results around a straight line (the regression line). At least two results are necessary: one to define the slope and one to define the intercept. Two degrees of freedom are thus "used up" in defining the position of the line and the number of degrees of freedom of the estimate of variance about the line is $N - 2$.

The principles just stated lead to a generalization concerning degrees of freedom; namely, the degrees of freedom always are the number of observations less the number of constants calculated from the data. The idea of degrees of freedom is fundamental to small-sample analysis, and it should be understood by those who would venture into the statistical thicket. If the number of observations is large, no great error is made if N is used instead of $N - 1$ in calculating the variance, and this is done often; but the multiple regression techniques and the tests of significance used in later sections require that the degrees of freedom be determined scrupulously. Also, a working knowledge of the idea of degrees of freedom often will help the engineer to decide whether his sample is adequate to do the job he expects of it.

The coefficient of variation is the standard deviation expressed as a percentage of the arithmetic mean, i.e.

$$v = 100s/\overline{X} \tag{1.13}$$

Its use is to illustrate the degree of spread in terms of the mean. It must not be used unless the origin of measurement is uniquely fixed. This precludes its use on measurements such as temperature which is related to an arbitrary origin.

Normal distribution

A special kind of frequency distribution that is very important in statistics is the *normal* or *Gaussian distribution*. The normal distribution is important because it forms the basis for most statistical measures of probability and error and it also is the basis of most sampling theory. Some statistical tests become unreliable when the data depart from normality, therefore transformations and special sampling procedures often are applied to get a normally distributed sample from a nonnormal population. Some of these will be discussed in later sections.

In appearance the normal distribution is a bell-shaped curve extending infinitely far in both positive and negative directions. The relative frequency with which a variable will take on values between two points is represented by the area under the curve between two points on the horizontal axis.

The generalized equation of a normal probability curve is

$$f(x) = (1/\sigma\sqrt{2\pi})e^{-1/2[(x-\mu)/\sigma]^2} \tag{1.14}$$

where

$f(x)$ = ordinate, height of curve
x = any value of the measured variable
μ = arithmetic mean of the population
σ = standard deviation of the distribution

Equation 1.14 refers to the entire population. A distinction must be made between the mean and standard deviation of a set of observed values—a sample—and the mean and standard deviation of the entire population. The observed mean and standard deviation can only be regarded as estimates of the true values, hence the proper notation for a sample is

$$f(x) = (1/s\sqrt{2\pi})e^{-1/2[(x-\overline{X})/s]^2} \tag{1.15}$$

where

s = standard deviation of the sample, Equation 1.10
$\overline{X}$ = sample mean, Equation 1.1

A normal probability curve can be "fitted" to sample data by writing the equation of, and drawing the graph of, a normal curve having the same arithmetic mean and standard deviation as the given data. Of course, the area under the curve must also be equal to the area under the histogram which is being fitted. It should be remembered that a normal curve can be fitted to any data, and the "fit" may be good or bad. A qualitative idea of the "goodness of fit" can be gained by superimposing the curve upon the histogram.

In fitting a normal curve to sample data, one must multiply Equation 1.15 by a factor, iN, to create a curve which will have the same area as the histogram. Then Equation 1.15 becomes

$$f(x) = (iN/s)(1/\sqrt{2\pi})\{e^{-1/2[(x-\overline{X})/s]^2}\} \tag{1.16}$$

where

i = length of class interval used to draw the histogram
N = number of observations

If the histogram is drawn on proportional frequency instead of actual frequency, N = 1.

Table 1F illustrates computation of a normal distribution curve, applying it to the example problem.

TABLE 1F

Bit No.	x = Footage drilled	$x - \overline{X}$	$(x - \overline{X})^2$
1	69	−32	1,024
2	72	−29	841
3	123	+22	484
4	135	+34	1,156
5	95	−6	36
6	89	−12	144
7	139	+38	1,444
8	109	+8	64
9	125	+24	576
10	80	−21	441
11	104	+3	9
12	53	−48	2,304
13	76	−25	625
14	102	+1	1
15	90	−11	121
16	116	+15	225
17	105	+4	16
18	108	+7	49
19	115	+14	196
20	110	+9	81
	2,015		9,837

$\overline{X} = 2{,}015/20 = 100.75$ ft
$s = \sqrt{9{,}837/19} = 22.8$ ft
$i = 20$ ft

x	p = $-\frac{1}{2}[(x - \overline{X})/s]^2$	e^p	f(x) = $(i/s\sqrt{2\pi})e^p$
40	−3.550	0.029	0.010
60	−1.597	0.203	0.071
80	−0.414	0.661	0.231
100.75	0.0	1.000	0.350
120	−0.356	0.701	0.245
140	−1.482	0.227	0.079
160	−3.377	0.034	0.012

Fig. 1.2 shows the smooth curve drawn through the computed f(x) values and superimposed upon the histogram. The f(x) curve forms the familiar bell shape which is typical of all normal distribution curves. The good fit between histogram and theoretical distribution curve indicates that the sample data are nearly normally distributed; and we may infer that the parent population probably is normally distributed. Of course, the fact that a small sample is normally distributed does not prove that the parent population is so distributed.

The mode, or most frequent value, is readily apparent as the value of

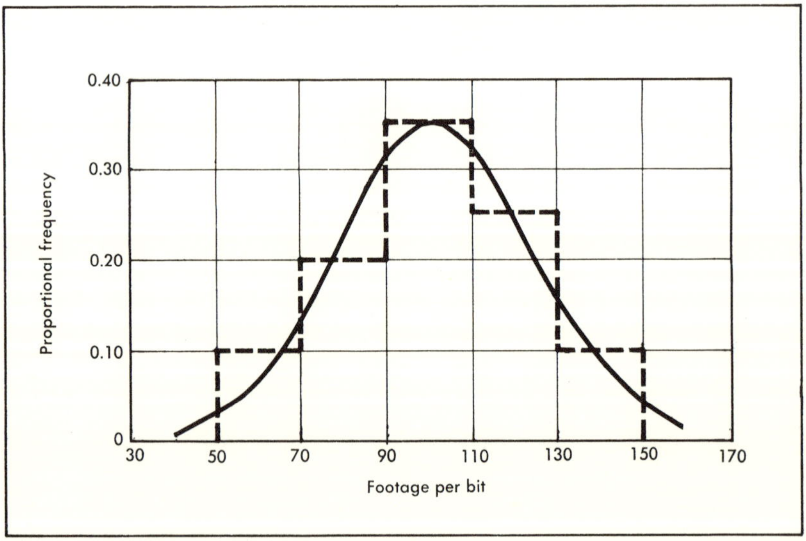

FIG. 1.2 Normal distribution function fitted to the bit-performance data of FIG. 1.1.

x which falls under the peak of the distribution curve—101 ft in the example. A characteristic of symmetrically distributed data is that the mode, median, and arithmetic mean have identical values.

With any probability density curve the area enclosed by verticals at x_1 and x_2 gives the *probability* that an observation will have a value between x_1 and x_2, and this probability multiplied by total number of observations is the expected number of observations in the range x_1 to x_2 for a sample of that size. For a normally distributed population, the area between x_1 and x_2 would be represented by an integration of the normal distribution curve between x_1 and x_2, i.e.

$$\text{Area} = \int_{x_1}^{x_2} (1/\sigma \sqrt{2\pi}) e^{-1/2[(x-\mu)/\sigma]^2}\, dx$$

Values of the integral as a function of $(x-\mu)/\sigma$ have been tabled in various statistical textbooks. A plot of the function is shown in Fig. 1.3. This is the cumulative normal distribution, and it gives the relative frequency with which observations will fall below (to the left of) any specified value.

The 20 measurements of the example problem do not constitute a large sample, but, for illustrative purposes, the distribution curve of Fig. 1.2 will be considered a probability density curve. If one should want to know the probability that an observed footage will have a value between 70 and 80, he can compute

$$(x_1 - \mu)/\sigma = (70 - 101)/22.8 = -1.36$$

and

$$(x_2 - \mu)/\sigma = (80 - 101)/22.8 = -0.92$$

and, consulting Fig. 1.3 he will find that the probability is $0.180 - 0.085 = 0.095$. There would be 0.095×20, or 2 bits that could be expected to attain footages of between 70 and 80 ft, according to the probability density curve.

A rule of thumb which is sometimes useful for estimating probabilities from a normal distribution curve is: one standard deviation encompasses 67% of the area under the curve, two standard deviations encompass 95% of the area and two and one-half standard deviations cover 99% of the area.

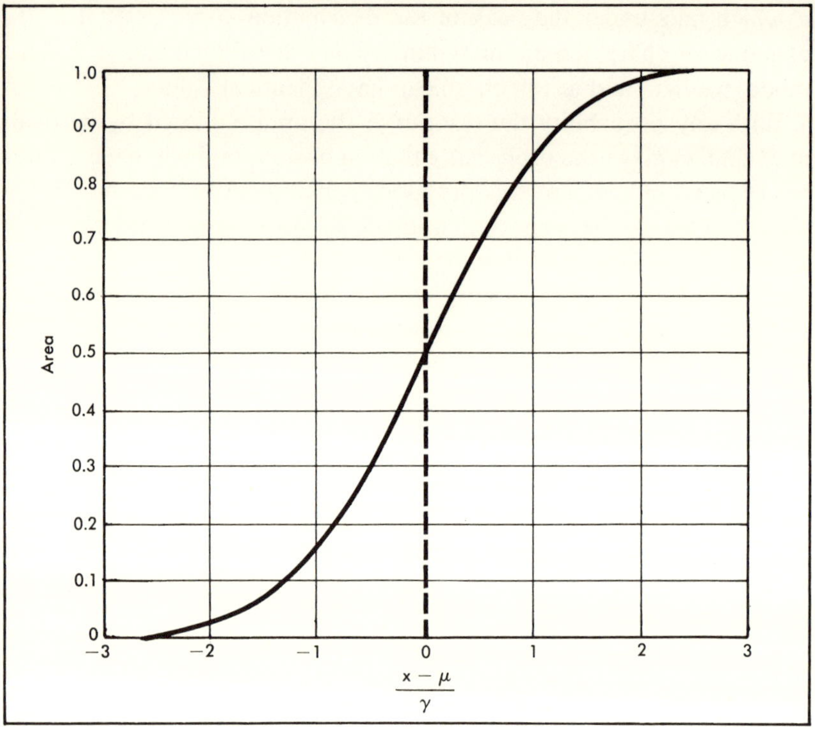

FIG. 1.3 The cumulative normal curve.

Nonnormal distributions

Most sample distributions are not perfectly symmetrical; the frequency may fall away more rapidly on one side of the maximum than on the other. This departure from symmetry is called *skewness,* and such distributions are said to be *skew.* Fig. 1.4 shows examples of skew distributions. Fig. 1.4a is skewed to the right and 1.4b is skewed to the left. In skew distributions the mode, median and mean no longer coincide as they do in normal distributions; instead, the median lies somewhere between the mode and mean. When probability density curves are skew, probabilities cannot be calculated from the cumulative normal curve of Fig. 1.3, but they can be estimated by the graphical method described later.

Occasionally systems will be encountered where the distribution exhibits more than one mode. A bimodal distribution is pictured in

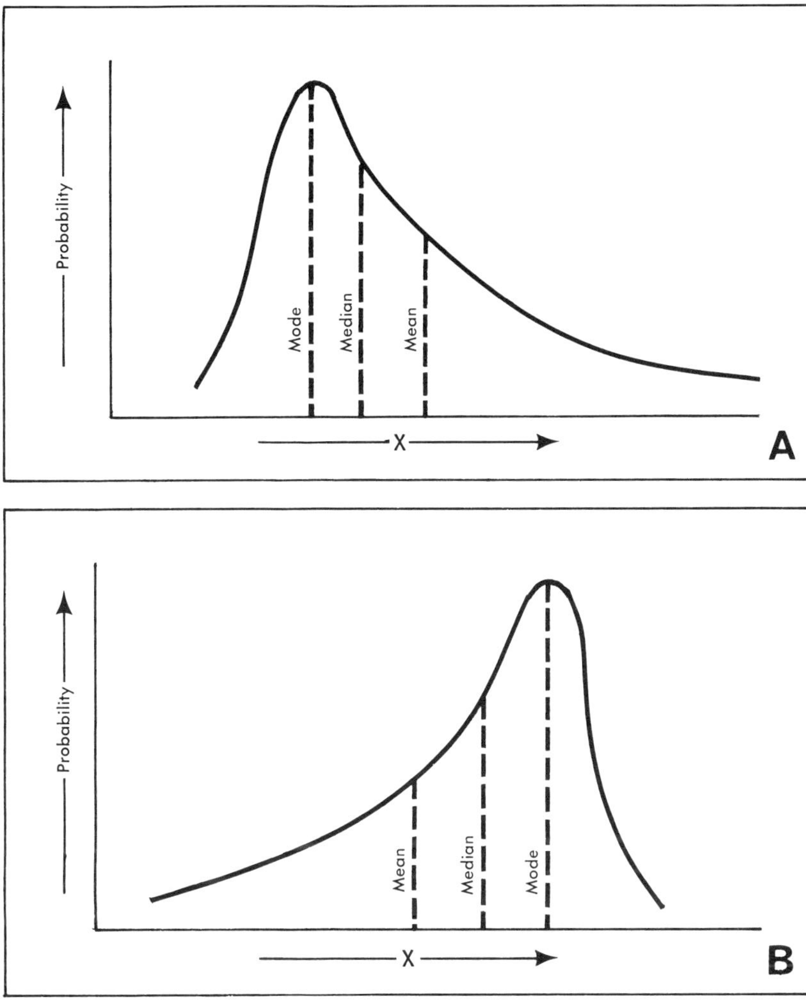

FIG. 1.4 Typical skew distributions.

Fig. 1.5. The presence of two or more modes usually means that the data are not homogeneous, that is, that two or more distinct distributions have been combined. The engineer should be careful when interpreting multimodal distributions as they could be caused by biased sampling or by the sample having been drawn from one or more discrete populations. Either situation can nullify the effectiveness of the sample for descriptive purposes.

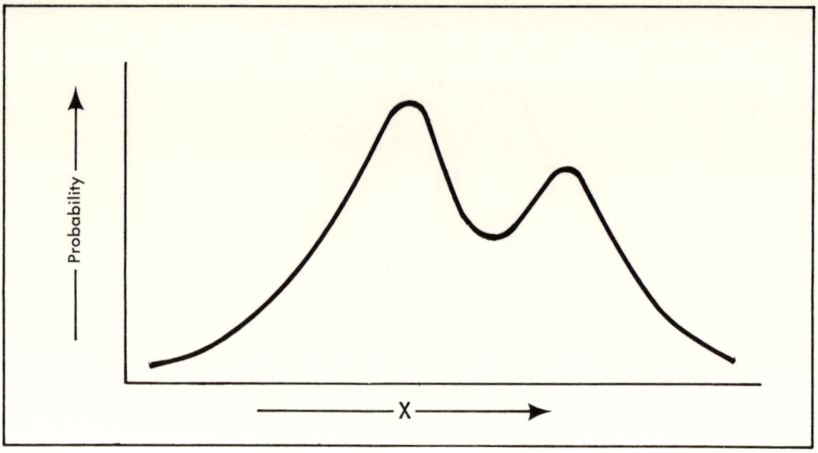

FIG. 1.5 A bimodal distribution.

Transformations

There are occasions when it is desirable to perform some kind of operation to transform the measured variable into a new variable. A transformation may be sought for one of the following reasons:

(1) A normal distribution may be created by transformation, making possible statistical tests which require normalcy.

(2) Some physical law may be better satisfied by the transformed variable than by the measured variable.

(3) A transformation may be made purely for aesthetic reasons e.g., to produce a symmetrical histogram instead of a skewed one.

Some transformations commonly used are log x, x^2, x^n, $\sqrt{x}$, $1/x^n$, log (x + a), and 1/(x + a). The exact transformation to use for a given type of data may require some trial and error to determine. A large number of observations usually is required before one can feel sure that his transformation is appropriate; thus it helps if the engineer is familiar with the type of data he is using.

A good example is Law's work with sandstone permeabilities. After studying a large amount of data he concluded that $\log_{\sqrt{2}}$ is the best transformation for sandstone permeability. Fig. 1.6 illustrates transformation of the permeabilities listed in the earlier problem. Fig. 1.6a is a histogram showing distribution of permeabilities and 1.6b is a histogram of $\log_{\sqrt{2}}$ K for the same data.

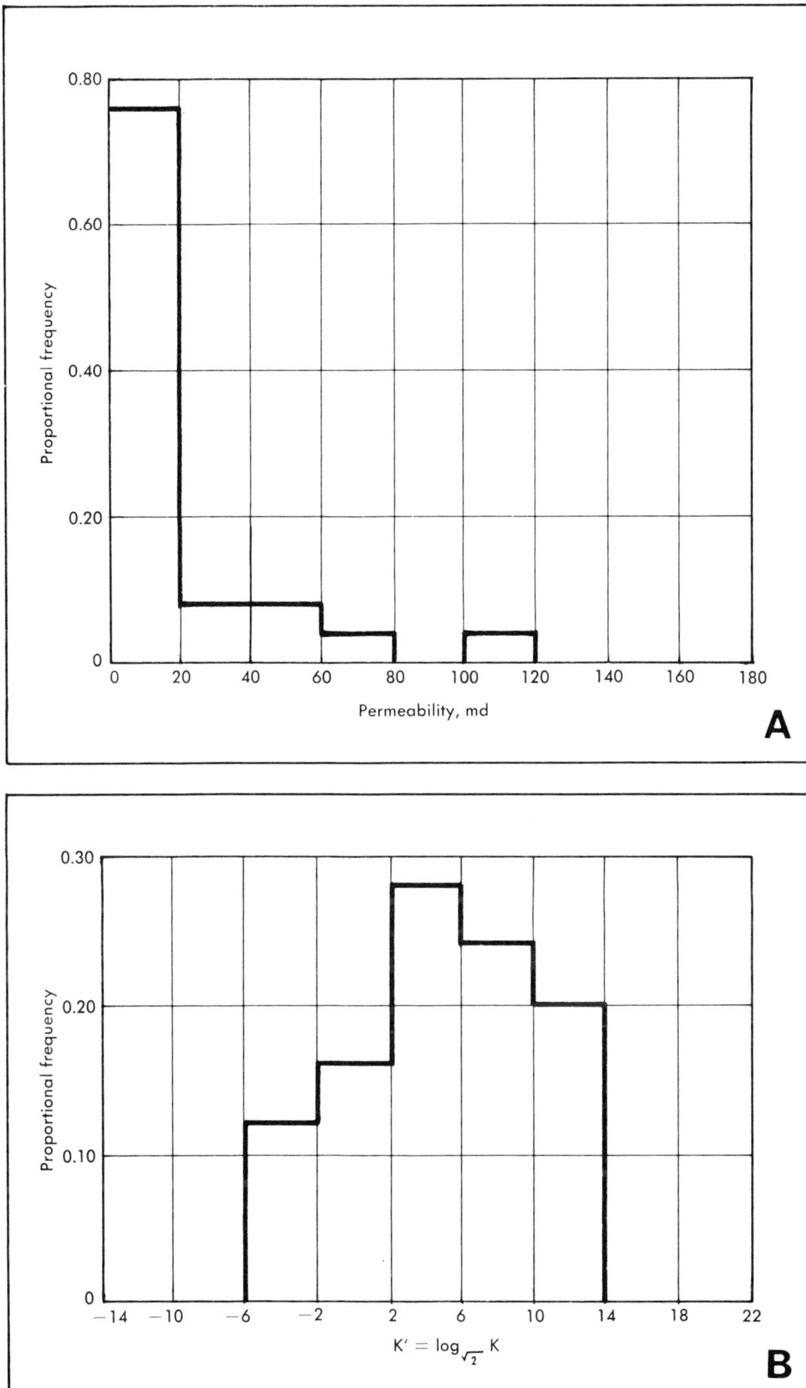

FIG. 1.6 Distribution of 25 permeabilities from a single well.

Coding

Manipulations that can be employed to simplify the computations involved in averaging large sets of numbers have been itemized in the section on "averages." These devices are so useful that it is worthwhile to summarize their effects in order that they can be remembered easily.

Adding (or subtracting) a constant from a group of observations will add (or subtract) the same constant to the arithmetic mean but will not change the variance or standard deviation.

Multiplying (or dividing) each observation by a constant will multiply (or divide) the arithmetic mean by the same constant, will multiply (or divide) the variance by the square of the constant, and will multiply (or divide) the standard deviation by the constant.

The process of adding or subtracting or multiplying or dividing by a constant is generally called *coding*.

Weighting

In a great majority of instances, data obtained may be considered to have equal validity. By this is meant, the error associated with making any particular measurement is believed to be the same, regardless of where or when the measurement is taken. If it were not true (for instance, if the error of measurement could vary with the measurement), then σ could not be taken as a constant of the population. If all the observations in a set have been affected by variable measurement error, they may be handled satisfactorily by a log transformation, but there may be times when the investigator has good reason to feel that only certain observations of the set contain unusual error. He would want to give less weight to these observations.

If some observations are worth w times as much as other observations, the problem can be handled by introduction of a weighting factor. For example, if a set of means are being averaged, one may have reason to want to count one of the means five times in order to give it more weight than the others. Then, the weighting factor would be $w = 5$. This particular mean would be worth $\sqrt{5}$ times as much as a value with a weighting factor of 1.0.

The weighted mean is given by Equation 1.6, and the weighted sum of squares is $\sum_{i=1}^{N} w_i x_i^2$. The error variance of $\overline{X}$ when there are weighted observations is

$$s_e^2 = \left[\sum_{i=1}^{N} w_i x_i^2 - \left(\sum_{i=1}^{N} w_i x_i\right)^2 / \sum w_i\right] / (w - 1) \qquad (1.17)$$

and the weighted crude sum of cross-products (useful in correlation analysis) is

$$\sum_{i=1}^{N} w_i x_i y_i - \sum_{i=1}^{N} w_i x_i \sum_{i=1}^{N} w_i y_i / \sum_{i=1}^{N} w_i \qquad (1.18)$$

As an example of coding and weighting, consider the problem shown in Table IG. Bottom-hole pressures have been measured in 14 wells, and all have been corrected to a common datum. Each well has been assigned a representative volume based on isopachs. The total reservoir volume represented by the 14 wells is 9,875 acre-ft. The representative volume is a weighting factor designed to represent the reservoir volume drained by each well.

Another type of weighting factor is shown as column 5 of Table 1G.

TABLE 1G

Well No.	Datum Pressure, P	Representative Acre-feet, V	P × V ÷ 1,000	Factor, F	F × P × V ÷ 1,000	F × V
1	2,910	775	2,255.25	2	4,510.50	1,550
2	2,875	660	1,897.50	2	3,795.00	1,320
3	2,928	360	1,054.08	2	2,108.16	720
4	2,883	890	2,565.87	2	5,131.74	1,780
5	2,907	925	2,688.98	2	5,377.96	1,850
6	2,813	330	928.29	1	928.29	330
7	2,867	780	2,236.26	2	4,472.52	1,560
8	2,895	940	2,721.30	2	5,442.60	1,880
9	2,912	835	2,431.52	2	4,863.04	1,670
10	2,925	435	1,272.38	2	2,544.76	870
11	2,826	600	1,695.60	1	1,695.60	600
12	2,803	560	1,569.68	1	1,569.68	560
13	2,900	875	2,537.50	2	5,075.00	1,750
14	2,905	910	2,643.55	2	5,287.10	1,820
Summation:	40,349	9,875	28,497.76	25	52,801.95	18,260

Arithmetic average $= \Sigma x_i/N = 40{,}349/14 = 2{,}882$

Volume-weighted average with equal weights for pressures

$= \Sigma w_i x_i / \Sigma w_i = (28{,}497.76/9{,}875) \times 1{,}000 = 2{,}886$

Volume-weighted average with unequal weights for pressures

$= \Sigma w_{i1} w_{i2} x_i / \Sigma w_{i1} w_{i2} = (52{,}801.95/18{,}260) \times 1{,}000 = 2{,}892$

This is a factor, arbitrarily assigned, to account for the fact that we believe some of the measurements in this particular set are not as reliable as the others. If one is going to do such a thing, he needs a reason, and the reason we have chosen not to give equal weight to wells 6, 11 and 12 is that we believe the pressures may have been measured in those wells before build-up was complete. We have given the pressure measurements from those wells one-half the value of measurements from the other wells, therefore the weighting factor of wells 6, 11 and 12 is 1.0, and the factor for all other wells is 2.0.

To introduce this idea into the data we simply count each measurement twice in all wells except 6, 11 and 12, which effectively gives us 25 measurements instead of 14. The example problem, then, contains two different types of weighting factors which have been applied for two very different reasons.

The pressures are to be averaged for use in material-balance calculations; and, though the variability between wells is rather large, experience shows that average values can be used effectively in most material-balance calculations if the averaging technique is precise and consistent.

The arithmetic average, or mean, is simply the sum of all pressures divided by the number of measurements, and its value is 2,882 psi. In computing the weighted averages, we have coded each pressure by dividing it by 1,000. This simply reduces the numbers to more manageable size.

In calculating weighted averages, one multiplies each measurement by its factor (or factors), sums these and divides the sum by the total of all the weighting factors. These computations are illustrated in Table 1G, where the volume-weighted average pressure is shown to be 2,886 psi, and the volume-weighted average with unequal weights for wells 6, 11 and 12 is 2,892 psi. The differences between the three averages may seem small in comparison to the differences between individual wells, but differences of this size can be significant in material balance calculations.

There is a certain variance and standard deviation associated with each of the averages just computed. Table 1H presents the elements for computing the variance and standard deviation on four averages from the pressure measurements of Table 1G. These are variance and standard deviation on: the arithmetic mean pressure, volume-weighted

TABLE 1H

Well No.	$P \div 1.000$	$P^2 \times 10^{-6}$	$V \div 1.000$	$PV \div 10^{-6}$	$(P \div 1.000)^2 \times (V \div 1.000)$	F	$F \times P \div 1.000$	$F \times V \times P \times 10^{-6}$	$F \times P^2 \times 10^{-6}$	$F \times V \times P^2 \times 10^{-9}$
1	2.910	8.4681	0.775	2.25525	6.56278	2	5.820	4.51050	16.9362	13.12556
2	2.875	8.2656	0.660	1.89750	5.45530	2	5.750	3.79500	16.5312	10.91063
3	2.928	8.5732	0.360	1.05408	3.08635	2	5.856	2.10816	17.1464	6.17269
4	2.883	8.3117	0.890	2.56587	7.39741	2	5.766	5.13174	16.6234	14.79481
5	2.907	8.4506	0.925	2.68898	7.81681	2	5.814	5.37796	16.9012	15.63370
6	2.813	7.9130	0.330	0.92829	2.61129	1	2.813	0.92829	7.9130	2.61128
7	2.867	8.2197	0.780	2.23626	6.41137	2	5.734	4.47252	16.4394	12.82272
8	2.895	8.3810	0.940	2.72130	7.87814	2	5.790	5.44260	16.7620	15.75633
9	2.912	8.4798	0.835	2.43152	7.08063	2	5.824	4.86304	16.9596	14.16117
10	2.925	8.5556	0.435	1.27238	3.72169	2	5.850	2.54476	17.1112	7.44339
11	2.826	7.9863	0.600	1.69560	4.79178	1	2.826	1.69560	7.9863	4.79177
12	2.803	7.8568	0.560	1.56968	4.39981	1	2.803	1.56968	7.8568	4.39981
13	2.900	8.4100	0.875	2.53750	7.35875	2	5.800	5.07500	16.8200	14.71750
14	2.905	8.4390	0.910	2.64355	7.67949	2	5.810	5.28710	16.8780	15.35903
Summation:	40.349	116.3104	9.875	28.49776	82.25159	25	72.256	52.80195	208.8647	152.70037

average pressure, arithmetic average pressure with unequal weights and volume-weighted average pressure with unequal weights.

The standard deviation on the arithmetic mean is a straightforward solution of Equation 1.11 and the variance is, of course, the square of Equation 1.11. The pressure has been divided by 1,000, therefore the variance must be multiplied by $1{,}000^2$ for the variance to be in correct units. The variance is

$$\begin{aligned}&\{[\Sigma \text{ col. } 3 - (\Sigma \text{ col. } 2)^2/N] \div (N - 1)\} \times 10^6\\&= \{[116.3104 - (40.349)^2/14] \div 13\} \times 10^6\\&= 1{,}669\end{aligned}$$

The standard deviation is $\sqrt{1{,}669} = \pm\ 40.85$ psi.

A special problem is encountered in computing the variance on the volume-weighted average pressure. Equation 1.17 is written for weighting factors which are integers, i.e., whole numbers. This is the type of weighting factor represented by F in the table. The volume-weighting factor is of a different type and it presents a problem when one starts to determine the degrees of freedom (w − 1) to enter into Equation 1.17.

In the example problem there are 14 weighting factors, they are all different and their sum is 9.875. We cannot deduct one measurement for degrees of freedom because we would not know which one to deduct, so the most reasonable thing to do is to deduct one average value, (9.875 ÷ 14) = 0.705. The variance on the volume-weighted average pressure with equal weights for all pressures then is

$$\begin{aligned}&\{[\Sigma \text{ col. } 6 - (\Sigma \text{ col. } 5)^2/\Sigma \text{ col. } 4] \div (\Sigma \text{ col. } 4 - 0.705)\} \times 10^2\\&= \{[82.25159 - (28.49776)^2/9.875] \div 9.170\} \times 10^6\\&= 1{,}238\end{aligned}$$

The standard deviation is $\sqrt{1{,}238} = \pm\ 35.19$ psi. Note that the representative volume, V, has been divided by 1,000 for computation purposes. Weighting factors of this type can be scaled in any way without affecting the calculated variance.

Looking at the other type of weighting device, we will see that Equation 1.17 applies directly. The variance on the arithmetic mean pressure with unequal weights for pressures is

$$\begin{aligned}&\{[\Sigma \text{ col. } 10 - (\Sigma \text{ col. } 8)^2/\Sigma \text{ col. } 7] \div (\Sigma \text{ col. } 7 - 1)\} \times 10^6\\&= \{[208.8647 - (72.2560)^2/25] \div 24\} \times 10^6\\&= 1{,}146\end{aligned}$$

The standard deviation is $\sqrt{1{,}146} = \pm\ 33.85$ psi.

When both types of weighting factors are used, the weighting factor for each measurement is $w_{i1} \times w_{i2}$ and the average weighting factor to be used in determining degrees of freedom is $\Sigma\ w_{i1}\ w_{i2}/N$. Then, in the example problem the average weighting factor is $\Sigma(\text{col. } 4 \times \text{col. } 7) \div N = 18.260/14 = 1.304$. The variance on the volume-weighted average pressure with unequal weights for pressures is

$$\begin{aligned} &\{[\Sigma \text{ col. } 11 - (\Sigma \text{ col. } 9)^2/\Sigma(\text{col. } 4 \times \text{col. } 7)] \\ &\qquad \div [\Sigma(\text{col. } 4 \times \text{col. } 7) - 1.304]\} \times 10^6 \\ &= \{[152.70037 - (52.80195)^2/18.260] \div (18.260 - 1.304)\} \times 10^6 \\ &= 849. \end{aligned}$$

The standard deviation is $\sqrt{849} = \pm\ 29.14$ psi.

In calculating variances it is essential that high precision be maintained during all the intermediate computations. Roundoff should be reserved for the final answer.

Perhaps a further word should be said about applying weighting factors to engineering measurements. The first type of factor, exemplified by the volume weighting factor in the example problem, is quite defensible and desirable; but the second type—the arbitrarily-assigned factor—should be used cautiously. It represents the introduction of a conscious bias into the results, and the engineer who applies it must be prepared to defend himself. Statistics must not be thought of as a license to capriciousness.

Graphical methods

One graphical device—the histogram—already has been illustrated. Perhaps a more descriptive presentation of data is afforded by use of normal probability graph paper. Normal probability paper has the horizontal axis scaled so that the cumulative probability curve of normally-distributed data will plot as a straight line. Of course, non-linearity of the plot would indicate that the distribution is not normal.

The cumulative normal probability plot has other uses in addition to providing a quick check for normality. Data which are nearly normal can be plotted and smoothed by eye. When the smoothed cumulative plot is differentiated graphically it produces a smoothed probability curve; this is a much less tedious procedure than fitting the curve to the data mathematically. The mode, median and probabilities can be approximated by reading them from the curve, thereby eliminating a lengthy least squares summation.

Because of the wide use of the $\log_{10}$ transformation, probability paper can be purchased which has a $\log_{10}$ scale along the vertical axis. This is for convenience rather than necessity because any transformation can be used with the linear paper provided the transformed variable is plotted instead of the original variable.

Table 1I shows the rock-bit data from the previous problem. The data have now been arranged as a cumulative probability tabulation—a form suitable for plotting on probability paper.

TABLE 1I

Footage per bit	No. of observations less than tabulated value	Percent of observations less than tabulated value
50	0	0
70	2	10
90	6	30
110	13	65
130	18	90
150	20	100

Fig. 1.7 is the cumulative probability plot. A straight line fits the sample distribution nicely, showing that it is approximately normal. The value at 50% probability (101 ft) is the median. Half of the footages lie below this value and half are above it. Since the distribution is symmetric (non-skew), 101 ft is also the mean and the mode. Percentages on the cumulative probability curve are referred to as *percentiles*. Thus, at the 80 percentile, 80% of the footages would be expected to be less than the observed footage.

The 16.7 and 83.3 percentile values of the variable are an approximation of the standard deviation limits. These are 78 ft and 123 ft, respectively. Likewise, the 95% probability interval is bounded by the 2.5 and 97.5 percentile values. These are 55 and 147 ft, respectively.

The following problem shows application of the graphical method to skewed data.

A high-permeability carbonate reservoir was being studied, and geological work showed that it was composed of at least two major petrologic types of carbonates. The question that needed to be answered

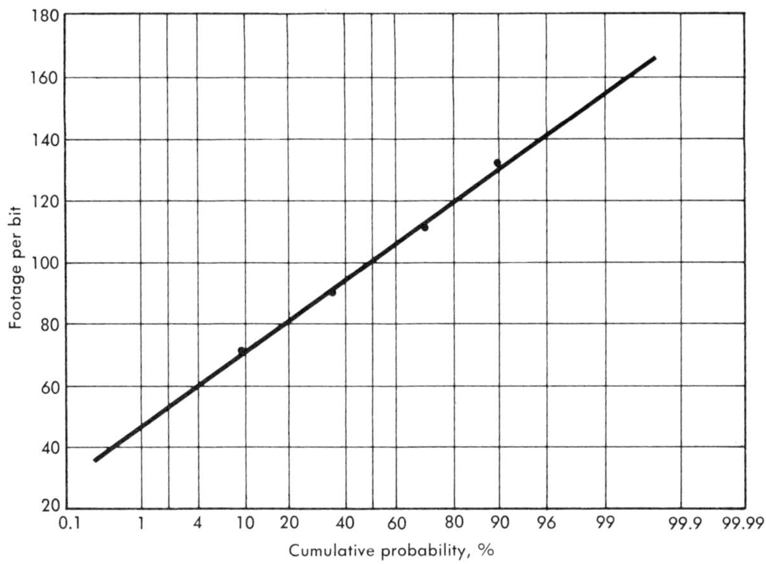

FIG. 1.7 Cumulative distribution of data from FIG. 1.1.

was whether these petrologic types really were two distinct "families" as far as pore structure was concerned. If they were distinctly different, it could be expected that they would have different recovery characteristics, therefore the reservoir should be zoned on a lithologic basis.

The pore characteristic selected for study was the mean pore radius. Pore-size distributions of 55 core plugs were calculated from mercury-injection capillary pressures by the method of Burdine, Gournay and Reichertz. The plugs covered the entire formation and were randomly selected from several wells. The mean pore radius is the weighted-average pore entry radius weighted on the basis of incremental pore volume.

It was expected that the distribution of mean pore radii would be bimodal if two distinct systems were represented in the sample. Table 1J shows the pore radii arranged as a cumulative distribution table. Fig. 1.8 is the cumulative probability plot of these data. The non-linearity of the plot is quite pronounced; obviously the distribution is skew. Also, there seems to be no bimodal tendency, but this would be more apparent from a probability curve.

TABLE 1J

R_{im}-microns	No. of observations less than tabulated value	Percent of observations less than tabulated value
0.4	0	
0.6	1	1.81
0.7	2	
0.7	3	
0.9	4	
1.0	5	9.10
1.3	6	
1.4	7	
1.5	8	
1.6	9	
1.7	10	18.20
1.9	11	
2.1	12	
2.3	13	
2.4	14	
2.5	15	27.23
2.5	16	
2.6	17	
3.1	18	
3.2	19	
3.2	20	36.30
3.6	21	
3.6	22	
3.7	23	
3.9	24	
3.9	25	45.35
4.1	26	
4.1	27	
4.1	28	
4.2	29	
5.0	30	54.50
5.2	31	
5.5	32	
5.5	33	
6.1	34	
6.1	35	63.60
6.5	36	
6.7	37	
6.8	38	
7.1	39	
7.5	40	72.75
7.8	41	
8.3	42	
8.5	43	
8.9	44	

TABLE 1J (*Continued*)

R_{im}-microns	No. of observations less than tabulated value	Percent of observations less than tabulated value
9.7	45	81.80
10.9	46	
11.0	47	
11.7	48	
13.3	49	
13.5	50	90.90
13.8	51	
14.1	52	
14.8	53	
21.8	54	98.25
26.7	55	

Differentiation of the smoothed cumulative curve is carried out in Table 1K.

Midpoint value plotted against probability—the distribution curve—is shown in Fig. 1.9. There is no bimodal tendency, but the distribution is badly skew. The most frequently-occurring pore radius is 2.5 microns, and half of the pore radii measured are less than 5 microns;

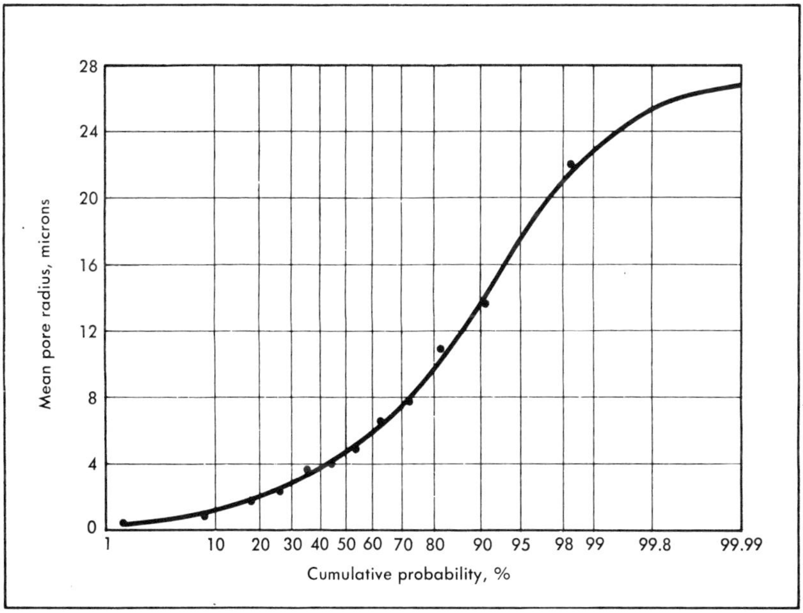

FIG. 1.8 Cumulative distribution of 55 mean pore radii.

TABLE 1K

Range	Midpoint value	Incremental probability
0–2	1	19.5
2–4	3	23.5
4–6	5	18.0
6–8	7	12.0
8–10	9	8.0
10–12	11	5.5
12–14	13	4.3
14–16	15	2.7
16–18	17	2.1
18–20	19	1.7
20–22	21	1.3

however, some are found to be as high as 20 microns. The skewness of the distribution is an indication of the heterogeneity of this formation and it can be expected to have its effect on flow characteristics.

Sampling

The term *sampling* is customarily thought of as the segregation of a small fraction from a large bulk of material in such a way that the characteristics of the bulk can be estimated by studying those of the sample

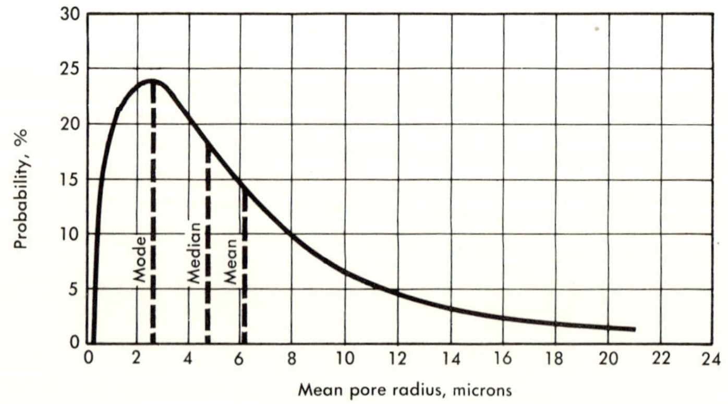

FIG. 1.9 Distribution of mean pore radii.

but, statistically speaking, any subset of a population is a *sample* from that population.

For example, if a core is cut from a formation, the core constitutes a sample of the formation and when 100 plugs are cut from the core each of these is a sample of the core. Then if the permeability of each plug is measured and the operator selects 10 of these permeabilities for averaging, these 10 permeabilities also are a sample. They are a sample of a set of measurements i.e., permeability.

It is unfortunate for the petroleum engineer that, more often than not, he has no control over the samples he has to use. Usually he must employ data that were measured by others, and quite often these measurements are made under unspecified conditions. Even when he gets the opportunity to make measurements himself he probably will find that economics or mechanics will keep him from controlling the tests as he would wish. It is obvious that his task will be more burdensome if he has reason to suspect that his data are not reliable.

If a sampling operation has any systematic error so that some individuals of the population are more likely to be chosen than others, the sample will be *biased*. Three of the most common causes of bias are:

(1) subjective sampling methods,
(2) calibration error in an instrument and
(3) failure of the operator to eliminate or control variables.

A commonly observed example of item (1) is the tendency—conscious or unconscious—of most samplers to show preference for the more permeable sections of a core when cutting plugs for analysis. As an example of item (2) suppose that of 30 gas meter installations in a field 10 have orifice plates whose edges are burred or worn at the orifice; then the average daily gas-oil ratio reported for this field will be too low. To illustrate item (3), if bottom-hole pressures are measured with a noncompensated bomb, the measured pressures will be in error and the error will depend upon temperature.

Items (2) and (3) fall into the category of engineering problems. Most engineers are thoroughly aware of them and make a conscientious effort to overcome them. But item (1) is a problem that may be less obvious to many.

The fact that the sampler can bias his sample unconsciously introduces a subtlety that makes it very important for him to adopt an objective system and stay with it.

The most universally-applicable technique for descriptive statistics is *random sampling*. A sample is randomly chosen when every individual in the population has an equal and independent chance of being chosen for a sample. Technically every individual chosen should be measured and returned to the population before another selection is made. This means, of course, that an individual could be chosen twice in the same sample. If the population is very large (or infinite as in many engineering problems), negligible error will result if the individual is not returned. But as an example of the other extreme, if one has only 30 measurements of permeability and he wants to study the distribution of the means of several random samples of 5 selected from these 30, he should be careful to retain the full set throughout his sampling.

While the subject of random sampling is being considered, it should be emphasized that the sampling boundaries, that is, the population limits, should be specified by the experimenter before he begins taking samples. This is because the sample could be biased if it is used to represent any other population, however similar it may seem to the one from which the sample is drawn.

A straightforward method of obtaining a random sample of N individuals from a finite population would be to assign a number to each member of the population, put a set of numbered tags corresponding to the individuals into a box and draw N tags from the box. This method may be quite satisfactory if the population is small, but for large populations it is preferable to use a table of random numbers. There are numerous tables of random numbers available in the literature, and there are many ways to use these tables. The tables are arranged so that the numbers 0 through 9 occur with approximately the same frequency. By combining the numbers in pairs one may get numbers from 00–99, by combining in threes he may get a set from 000 to 999, etc. The simplest practice is to enter the table randomly and go down the column picking numbers in succession, rejecting those that are too large.

It should be noted that subjective sampling is not always undesirable, but it puts the burden on the engineer and not the statistician. For example, in the case of the rock bit data, it may be suggested that the engineer should reject any measurements taken under ambiguous circumstances. In this problem he had no control over the test conditions but he did have control over the sampling; however, he could justify arbitrary rejection only in measurements where he knew there was ambiguity and could point to its cause. In core work the sampling procedure might be random or subjective, depending on the use to which the sample will be put. If the purpose of sampling is to describe the heterogeneity of a section, random sampling (e.g. plugging on a fixed interval, regardless of lithology) should be used; but if the engineer or geologist is familiar with the formation and wants an average from the most productive section, he may concentrate on taking plugs from that section.

OIL INDUSTRY REFERENCES

Bennion, D. W. and Griffiths, J. C.: "A Stochastic Model for Predicting Variations in Reservoir Rock Properties," Trans. AIME (1966), Vol. 237, p. II-9.

Buckner, H. P., Jr., Felsenthal, M. and Conley, F. R.: "A Simplified Pore Size Distribution Apparatus," Trans., AIME (1956), Vol. 207, p. 306.

Bulnes, A. C.: "An Application of Statistical Methods to Core Analysis of Dolomite Reservoirs," Trans., AIME (1946), Vol. 165, p. 223.

Burdine, N. T., Gournay, L. S. and Reichertz, P. P.: "Pore Size Distribution of Petroleum Reservoir Rocks," Trans., AIME (1950), Vol. 189, p. 195.

Burdine, N. T.: "Relative Permeability Calculations from Pore Size Distribution Data," Trans., AIME (1953), Vol. 198, p. 71.

Brownscombe, E. R. and Collins, Francis: "Estimation of Reserves and Water Drive from Pressure and Production History," Trans., AIME (1949), Vol. 186, p. 92.

Cardwell, W. T., Jr. and Parsons, R. L.: "Average Permeabilities of Heterogeneous Oil Sands," Trans., AIME (1945), Vol. 160, p. 34.

Dotson, B. J., Slobod, R. L., McCreery, P. N., and Spurlock, James W.: "Porosity-Measurement Comparisons by Five Laboratories," Trans., AIME (1951), Vol. 192, p. 341.

Dykstra, H. and Parsons, R. L.: "The Prediction of Oil Recovery by Water Flood," Secondary Recovery of Oil in the United States, API (1950).

Gatlin, Carl and Nemir, Charles E.: "Some Effects of Size Distribution on Particle Bridging in Lost Circulation and Filtration Tests," Journal of Petroleum Technology (June, 1961), p. 575.

Greenkorn, R. A. and Johnson, C. R.: "Variation of a Natural Sandstone Reservoir Element: An Objective Analysis of Core Measurements," Paper No. 1577G presented at 35th Annual Fall Meeting of SPE in Denver (October, 1960).

Griffiths, J. C.: "Grain-size Distribution and Reservoir Rock Characteristics," Bull., AAPG (Feb., 1952), Vol. 36, p. 205.

Haring, R. E., Adams, R. B., Beazley, R. A. and Kipp, K. L.: "Application of Probability to the Design of Single-Point Moorings, "Trans., AIME (1970), Vol. 249, p. II-219.

Inman, D. L. and Chamberlain, T. K.: "Particle Size Distribution in Nearshore Sediments," Finding Ancient Shorelines, SEPM Special Publication No. 3.

Johnson, P. O., "Statistical Methods in Research," p. 18.

Katz, D. L., Turner, C. E., Grimm, P. D., Elenbaas, J. R. and Vary, J. A.: "Sample Grading Method of Estimating Gas Reserves," Trans., AIME (1952), Vol. 195, p. 207.

Katz, Marvin L. and Tek, M. Rasin: "A Theoretical Study of Pressure Distribution and Fluid Flux in Bounded Stratified Porous Systems with Crossflow," Paper No. 146 presented at Fall Meeting of SPE (October, 1961).

Kendall, V. V.: "Statistical Analysis of Test Containers for Condensate Well Corrosion Studies," Oil and Gas Journal, NACE Abstracts (April 19, 1947), p. 67.

Law, Jan: "A Statistical Approach to the Interstitial Heterogeneity of Sand Reservoirs," Trans., AIME (1944), Vol. 155, p. 202.

Law, Jan, and Cerini, William F.: "A Statistical Approach to Core Analysis Interpretation," Drilling and Production Practices, API (1946), p. 373.

Maly, G. P. and Krueger, R. F.: "Improper Formation Sampling Leads to Improper Selection of Gravel Size," *Journal of Petroleum Technology* (December, 1971), p. 1403.

Otto, G. H.: "A Modified Logarithmic Probability Graph for the Interpretation of Mechanical Analysis of Sediments," Journal of Sedimentary Petrology (August, 1939), Vol. 9, p. 62.

Rogers, E. B.: "Sand Control in Oil and Gas Wells—1," *Oil and Gas Journal,* (Nov. 1, 1971), p. 54.

Stanley, L. T.: "Curve-Fitting Cuts Material Balance Calculations," Petroleum Engineer (August, 1961), p. 116.

Stiles, W. E.: "Use of Permeability Distribution in Water Flood Calculations," Trans., AIME (1949), Vol. 186, p. 9.

Warren, J. E., Skiba, F. F. and Price, H. S.: "An Evaluation of the Significance of Permeability Measurements," Journal of Petroleum Technology (August, 1961), p. 739.

Warren, J. E. and Price, H. S.: "Flow in Heterogeneous Porous Media," SPE Journal (September, 1961), Vol. 1, p. 153.

2

Confidence limits and tests of significance

Introduction to statistical tests

Frequency distributions were discussed at some length in Chapter 1. In that discussion it was stated that most of the important statistical tests are based upon normal distribution. Also, it was pointed out that populations often are not normally distributed and that the distribution of a sample does not necessarily reflect the distribution of the parent population. There are some tests that are more *robust* than others (the t-test is one of these); that is, they are relatively unaffected by departure from normality. Nevertheless, any of the tests will be weakened to some degree by departure of the data from a normal distribution.

It may seem, then, that statistical tests have little practical value, but such is not the case. This is because most of the important tests work with distributions of sample means; and statisticians have shown that *the means of small, randomly-selected samples approach normal distribution even if the raw data are not normally distributed.*

It can also be shown that the standard deviation of the means of samples drawn randomly from any population will approach a value equal to $\sigma/\sqrt{N}$, where σ is the standard deviation of the population and N is the number of observations in each sample. The statistic, $\sigma/\sqrt{N}$, is called *the standard error of the mean.* The standard error of the mean is a *statistic* that recurs often in the arithmetic of statistics.

Two important properties of the standard error of the mean are:

(1) The reliability of a mean is related to the sample size. The standard error is inversely proportional to the square root of the sample

size, so that if the sample size is increased four times, the standard error of the mean will be halved.

(2) Whatever the nature of the parent population, the distribution of sample means tends rapidly to the normal form as sample size increases. If the parent population is roughly normal, samples containing as few as four observations may be satisfactory.

Following is a small sampling experiment which illustrates how the sampling error of the mean can be estimated, even in non-normal populations, from the means of small, randomly selected samples.

There are permeabilities measured on 277 core plugs from an oil reservoir. No values less than 1 md were reported, because rock having permeability in this range was considered nonproductive. The arithmetic mean permeability is 23.4 md and the standard deviation is 40.4 md. These 277 permeabilities represent the entire body of information concerning the permeability of this reservoir; so, for the sake of illustration, they will be considered as constituting the total population. Then, $\mu = 23.4$ md, and $\sigma = 40.4$ md.

Remember that in the terminology of statistics, μ is the true mean of the population and σ is the true standard deviation of the population distribution. These are to be distinguished from $\overline{X}$, a sample mean, and s, the standard deviation of a sample distribution. The statistics $\overline{X}$ and s are simply estimates of the true values.

Twenty samples, each containing eight observations, were drawn randomly (a table of random numbers was used according to the manner described in Chapter 1) from the list. The results are shown in Table 2A. The sample means range from 5.4 to 53.4 md.

This is an appreciably smaller range than the 1.0 to 279.0 md of the parent population. The overall sample mean, $\overline{K} = 564.7/20 = 28.23$ md compares favorably with the population mean, $\mu_k = 23.4$ md. We would expect the standard deviation of the sampling means, $s_{\overline{K}}$, to be an approximation of the standard error of the mean. It turns out that $s_{\overline{K}} = \sqrt{[\Sigma\overline{K}^2 - (\Sigma\overline{K})^2/N]/(N-1)} = 15.4$ md, whereas $\sigma/\sqrt{N} = 40.4/\sqrt{8}$ $= 14.3$ md.

If the distribution of $\overline{k}$'s is plotted on cumulative probability paper by the method described in Chapter 1, the distribution will be seen to trend toward normal, but irregularities of the curve will be seen. Considerable improvement in the approach to normality probably could be achieved by using a sample size larger than eight.

TABLE 2A

Sample No.	k_i – Observed permeability – md								Σk_i	$\bar{k}$	$\bar{k}^2$
1	16.0	89.0	5.6	1.3	7.4	167.0	6.8	1.4	294.5	36.8	1354.2
2	15.0	19.0	1.8	1.5	1.1	6.2	6.1	42.0	92.7	11.6	134.6
3	4.7	4.3	54.0	16.0	89.0	47.0	45.0	3.2	263.2	32.9	1082.4
4	5.8	1.9	32.0	53.0	5.7	2.6	7.9	8.8	117.7	14.7	216.1
5	78.0	45.0	17.0	3.4	3.9	28.0	149.0	43.0	367.3	45.9	2106.8
6	182.0	3.8	75.0	4.6	56.0	1.2	21.0	1.6	345.2	43.2	1866.2
7	60.0	6.8	112.0	34.0	57.0	1.4	45.0	1.0	317.2	39.6	1568.2
8	1.0	31.0	57.0	128.0	16.0	182.0	5.3	1.0	421.3	52.7	2777.3
9	3.9	4.6	17.0	1.6	15.0	5.8	121.0	157.0	325.9	40.7	1656.5
10	1.5	1.8	54.0	12.0	24.0	6.0	58.0	11.0	168.3	21.0	441.0
11	3.9	51.0	1.2	15.0	3.6	2.2	1.9	3.3	82.1	10.3	106.1
12	6.0	2.0	7.9	2.2	5.8	1.4	2.7	15.0	43.0	5.4	29.2
13	41.0	30.0	1.0	2.5	101.0	1.9	12.0	6.0	195.4	24.4	595.4
14	2.2	55.0	10.0	2.1	1.9	4.5	1.9	17.0	94.6	11.8	139.2
15	19.0	5.7	3.5	8.5	19.0	11.0	31.0	2.4	100.1	12.5	156.3
16	57.0	2.2	4.3	2.6	3.4	2.7	11.0	1.2	84.4	10.6	112.4
17	120.0	5.4	11.0	5.1	120.0	21.0	4.6	19.0	306.1	38.3	1466.9
18	3.4	54.0	3.8	47.0	13.0	19.0	13.0	45.0	198.2	24.8	615.0
19	1.9	58.0	19.0	9.4	60.0	3.8	1.0	120.0	273.1	34.1	1162.8
20	167.0	1.0	1.9	16.0	13.0	54.0	149.0	25.0	426.9	53.4	2851.6
									4517.2	564.7	20438.1

$$\bar{\bar{k}} = 564.7/20 = 28.23 \text{ md}$$

$$s_{\bar{k}} = \{[20{,}438.1 - (564.7)^2/20]/19\}^{1/2} = \sqrt{236.5} = 15.4 \text{ md}$$

This problem was presented only for the purpose of illustrating the principle that the distribution of sample means tends to normality, even though the distribution of the parent population is not normal (skew). It is not necessary for the engineer to carry out the kind of tedious sampling procedure described above, because statisticians, by application of the mathematical theory of probability, have produced tables of theoretical sampling distributions and standard error of a variety of statistical parameters. The engineer can test his sample-derived values against the theoretical values and assess the reliability of his sample. These procedures comprise the tests of significance and measurement of confidence limits which are to be described shortly.

Special techniques have been devised by statisticians for handling unusually small samples, estimates based on inefficient statistics, distribution-free approximations, etc. These are of limited interest to the engineer and will not be discussed here. Examples of their application are offered in most statistical texts.

Confidence limits

When a statistic is calculated from a limited sample, the result is subject to uncertainty. The degree of uncertainty can be expressed by a confidence interval within which the true value is almost certainly included. Confidence intervals can be determined for estimates of the mean or estimates of the standard deviation. In the case of means a close approximation is obtained whether or not the data are normally distributed, but with the standard deviation a departure from normality may affect the confidence limits appreciably.

Statisticians have shown that when most populations are sampled randomly, the ratio,

(sample mean − true mean) ÷ (standard deviation of sample means)

is normally distributed and has zero mean and unit standard deviation. If the true mean is specified as μ_o, and the population variance is σ^2, then the ratio is

$$Z = (\overline{X} - \mu_o) \;/\; (\sigma / \sqrt{N}) \tag{2.1}$$

The mean value of Z is zero *only* if μ_o really is the true population mean. If one were to observe a large number of values of sample mean,

$\overline{X}$, which caused Z to be very much different from zero, one would question the hypothesis that μ_o is the true mean.

The statistic Z is difficult to evaluate because true standard deviation, σ, usually is not known; so σ must be replaced by sampling deviation, s, and

$$t = (\overline{X} - \mu) / (s / \sqrt{N}) \quad (2.2)$$

is used instead of Z. The sampling distribution of t is the generalized statistic that is used in assigning confidence limits to sample means. It is also used in a test of significance called the t-test.

Even though on the average s^2 is equal to σ^2, s^2 is actually less than o^2 more than half the time. This has been shown by experimentation. If N is very large and μ is known with certainty, s^2 will be very close to σ^2 and the distribution of t will differ little from the distribution of Z; but as N gets smaller, the difference between sampling distribution t and true distribution Z gets larger. Since the t distribution depends upon the size of N, separate curves have been derived at degrees of freedom ranging from 1 to ∞. The data from these curves are arranged in the form of a table (Table A-1, appendix A). The table gives the expected value of t at a given number of degrees of freedom and a given probability level. The t table represents a two-sided distribution; i.e. the distribution of t is a symmetrical probability curve, and at any probability level the numerical value is both plus and minus. This means, for example, that the 95% probability interval would be bounded by $t_{.975}$ and $-t_{.025}$. (Interpolation is permissible within Table A-1 and the other statistical tables, but one should observe the restrictions contained in footnotes to certain of the tables).

A fraction of the observations of sample mean, $\overline{X}$, will fall outside the range of whatever confidence interval may be selected. This fraction is represented by the symbol, α.

If the probability level is specified as $1 - \alpha$ and equation 2.2 is rearranged,

$$\mu = \overline{X} \pm t_{1/2\alpha} s / \sqrt{N} \quad (2.3)$$

This states that there is a probability $1 - \alpha$ that the true population mean will lie somewhere between $\overline{X} - t_{1/2\alpha}\ s/\sqrt{N}$ and $\overline{X} + t_{1-1/2\alpha}$ $s/\sqrt{N}$. The probability is known as the *confidence level* and the two solutions of Equation 2.3 are called confidence limits. For an example of *confidence limits,* please refer to the sampling data of Table 2A. If we were to consider sample 1, the only information available concerning the population, $\overline{K}_1 = 36.8$ md, would be our best estimate of the population mean, and $s_1 = \sqrt{[\Sigma K_i^2 - (\Sigma K_i)^2/N]/(N - 1)} = \sqrt{[36{,}202 - (294.5)^2/8]/7} = 60.2$, our best estimate of the population standard deviation. With this information we could place confidence limits on the estimate of the mean, using Equation 2.3. If we chose to work at the 95% probability level, we would enter Table A-1 at $1 - 1/2\ \alpha = 0.975$ and $N - 1 = 7$ degrees of freedom. There we would find $t = 2.365$. Then, $\mu = 36.8 \pm (2.365)(60.2/\sqrt{8})$, making the 95% confidence limits on the estimate of μ equal to -13.5 and 87.1. What these limits tell us is that 95% of the limits obtained in this way will include the population mean. In other words, there is a 5% chance that this particular interval does not include the population mean.

The wide interval set by the confidence limits of a sample with only eight measurements would probably convince us that a larger sample is needed. If we were to make 40 more measurements at random, this would bring the sample size to 48. This new sample would include samples 1 through 6 in Table 2A. The new population estimates would be $\overline{K} = 30.8$ md and $s = 43.3$ md, and the new confidence limits would be $30.8 \pm (2.014)(43.3/\sqrt{48}) = 18.2$ and 43.4. These limits indicate that the estimate of μ has been considerably improved by use of a larger sample. Note that the theoretical value of t, which is 2.014, was obtained by interpolation in Table A-1 at 47 degrees of freedom.

The example just cited is unusual because the population mean was specified. In an actual reservoir problem the true mean permeability of the reservoir would never be known.

Confidence limits need not be restricted to estimates of the mean. They can be applied to estimates of the standard deviation, but (unlike the confidence limits on the mean) they are highly sensitive to departures from normality. The method described below is a distribution-free method that gives approximate confidence limits on the standard deviation *whatever the distribution of the population.* The method requires a greater number of samples than most statistical tests, but in the relatively rare instances when an engineer will be interested in confidence limits on an estimate of σ, the additional sampling probably will be amply justified by the advantage of a distribution-free method. Basic data should consist of at least 30 observations for the method to be reliable. The procedure is:

(1) Divide the observations at random into equal groups, by using random numbers or by drawing numbered slips of paper. (This is exactly the same procedure followed in the sampling experiment, Table 2A). The number of observations in each group will depend upon the data. In general, if the original data are highly nonnormal, a greater number of observations in each group will be required than if the data are nearly normally distributed.

(2) Calculate the variance, s^2, for each of the groups and take the logarithms of the variances.

(3) Treating the transformed variances as new variables, calculate the mean and standard deviation.

(4) Apply the t-distribution to calculate the confidence limits in the same manner used for the confidence limits on a mean.

The sampling data of Table 2A will be used to illustrate the confidence limits on a variance. Because the distribution of the parent population is badly skewed (as usually is true with permeability data), it will be desirable to have a relatively large number of observations in each group. If we combine the samples of Table 2A into pairs, e.g. samples 1 and 2, 3 and 4, etc., we will have 10 groups of 16 observations each. This should provide sufficient information for a good estimate of the confidence limits on σ.

Table 2B, below, is a tabulation of the variance estimates computed from the data of Table 2A.

TABLE 2B

Group	$s_{Ki}^2 = [\Sigma K_i^2 - (\Sigma K_i)^2/16]/15$	$V' = \log (s_{Ki}^2)$	$(V')^2$
1	1,950.92	3.290	10.824
2	690.62	2.840	8.066
3	2,926.39	3.466	12.013
4	2,836.1	3.452	11.916
5	2,118.3	3.326	11.062
6	151.46	2.181	4.757
7	744.49	2.872	8.248
8	214.64	2.332	5.438
9	1,451.96	3.162	9.998
10	3,017.87	3.480	12.110
		30.401	94.432

The estimated mean and standard deviation are computed in the usual way.

Estimated mean of log variance = 30.401/10 = 3.0401

Estimated standard deviation of log variance

$= \sqrt{[\Sigma(V')^2 - (\Sigma V')^2/N]/(N-1)} = \sqrt{[94.432 - (30.401)^2/10]/9}$
$= 0.473$

95% Confidence limits on log of estimated mean variance

$$= 3.0401 \pm (2.262)(0.473/\sqrt{10})$$
$$= 2.702 \text{ and } 3.378$$

Estimate of σ^2 = antilog 3.0401 = 1097

95% Confidence limits on estimate of σ^2

$$= \text{antilog } 2.702 \text{ and antilog } 3.378$$
$$= 503 \text{ and } 2385$$

Estimate of $\sigma = \sqrt{1097} = 33.1$ md

95% Confidence limits on estimate of σ

$$= \sqrt{503} \text{ and } \sqrt{2385} = 22.4 \text{ md and } 48.8 \text{ md}$$

It is customary for investigators to work at the 95% confidence level, but the confidence level selected will depend upon the type of data and the degree of certainty desired by the investigator.

Tests of significance – the t-test

Tests of significance are closely related to confidence limits. In placing confidence limits on a sample mean, the investigator selects a probability level, reads t from the table and uses it to calculate the confidence limits, but in making the t-test he computes t from the data and compares it to the tabled values of t to find the probability level. The t-test is used in comparing two means, its purpose being to tell us whether the means are significantly different. The probability level corresponding to computed t is the probability that the two means are truly different. For example, assume a sample consisting of 7 measurements produces a value, t = 1.6. Entering Table A-1 at 6 degrees of freedom, we find that t = 1.6 falls in the table at $t_{.92}$. This indicates that in 16% of samples of this type a value of t as large as 1.6 could be

expected to occur by chance when, in fact, t is less than 1.6. Another way of stating it is that there is a 16% probability that a sample "t" of 1.6 is not significantly different from "t" in the table.

In testing the significance of means, two means are compared and the t-test provides an objective method for determining whether the two are significantly different. *The conventional procedure is to declare the difference significant at the 95% probability level and highly significant at the 99% level.* A single asterisk superscript signifies that the t is significant and a double asterisk that it is highly significant.

The t-test consists of testing the truth of a *hypothesis that two means are the same*. Such a hypothesis is called a *null hypothesis*. The procedure is to calculate the probability of finding a difference on the assumption that the null hypothesis is true. If this probability is sufficiently small, the null hypothesis is rejected and we conclude that the two means are truly different. Some types of comparisons that can be made between two sample means are illustrated by the examples following:

As an example, suppose that corrosion rate is being measured in a flow-line system by use of corrosion coupons installed at a number of different places in the system. Without a corrosion inhibitor the average corrosion rate was found to be $\overline{C}_1$ mils per year (mpy). A certain corrosion inhibitor was tested for an extended period of time, and the average corrosion rate of the test coupons was found to be $\overline{C}_2$ mpy. It happens that $\overline{C}_2$ is less than $\overline{C}_1$, and we would like to know whether the corrosion rate was actually reduced by the inhibitor or whether the two means, $\overline{C}_1$ and $\overline{C}_2$, are different simply because of sampling variations.

We will assume that the standard deviation, σ, is the same for both means, because average line temperature and composition of the flowing fluids was similar in both tests. The null hypothesis to be tested is that $\overline{C}_1$ and $\overline{C}_2$ actually are not significantly different.

In this case we are assuming that $\overline{C}_1$ represents the true mean, μ_1. It is possible, by using Table A-1, to calculate how small $\overline{C}_2$ is likely to be if the true mean is μ_1.

For example, it is unlikely (1 chance in 40) to be less than $\mu_1 - 1.96\ \sigma/\sqrt{N}$ where 1.96 is t of an infinite population, σ is the standard deviation of that population, and N is the number of coupons comprising the inhibitor test. If $\overline{C}_2$ is less than $\mu_1 - 1.96\ \sigma/\sqrt{N}$, we will reject the null hypothesis and conclude that the corrosion inhibitor has caused a reduction in corrosion rate.

This is a *one-sided test,* because we are interested only in whether $\overline{C}_2$ is significantly less than $\overline{C}_1$. In a *two-sided test* the null hypothesis would take the form $\mu \pm t\, \sigma/\sqrt{N}$ and the significance level would be 2α.

Example 1

The mol percent of carbon dioxide in the gas from four wells was measured at 25.80%, 22.17%, 23.69% and 24.78%. What is the probability that the true carbon dioxide content is 25.00%?

$$\mu = 25.00$$
$$\overline{X} = (25.80 + 22.17 + 23.69 + 24.78)/4 = 24.11$$
$$s = \sqrt{[\Sigma x^2 - (\Sigma x)^2/N]/(N-1)} = [(2332.41 - 9300.67/4)/3]^{1/2}$$
$$= \pm 1.55\%$$
$$t = (\mu - \overline{X})/(s/\sqrt{N}) = (25.00 - 24.11)/(1.55/2)$$
$$= 1.16$$

The degrees of freedom are N − 1, because one degree of freedom was used in computing the mean, therefore df = 3. Entering Table A-1 at df = 3, we see that the sample t—by interpolation—corresponds to $t_{0.83}$ in the table; therefore $1 - \alpha$ is $1 - 0.83 = 0.17$. This will be a two-sided test, because for all we know $\overline{X}$ can be larger or smaller than the postulated mean of 25.00 in any particular sample we may draw. Being a two-sided test the probability level is $1 - 2\alpha = 1 - 2 \times 0.17 = 0.66$. The implication of this result is that the true mean could be 25.00 and a value of 24.11 could occur 34% of the time by chance alone. Another way of stating it is that the sample mean of 24.11 does not vary significantly from a true mean of 25.00.

Example 2

The solution gas-oil ratio of an oil measured in the laboratory by flashing a bottom-hole sample through simulated separation stages is 275 scf/bbl. Gas-oil ratios in the field were measured at the separators of 10 wells and they are as follows: 280, 241, 309, 321, 275, 308, 327, 243, 311 and 335. There is variability from well to well because of inaccuracies of metering and because the different separators are not operating at exactly the same conditions. The average field gas-oil ratio is 295 scf/bbl. Is this a significant increase in GOR, indicating that free gas flow has begun in the reservoir?

$$\mu = 275;\ \overline{X} = 295;\ df = 9$$
$$s = [(880{,}416 - 8{,}702{,}500/10)/9]^{1/2} = \pm 33.6$$
$$t = (295 - 275)/(33.6/3.16) = 1.906$$

The calculated t corresponds to t in Table A-1 at a probability of 91% on a two-sided test; therefore, the hypothesis that the true mean is 275 cannot be rejected at the 95% probability level, and it must be concluded that the 295 figure does not represent a significant change in the gas-oil ratio.

Example 3

Core analyses from two wells show the following porosities:

Well 1	Well 2
1.7	5.9
5.9	6.9
1.5	3.6
4.1	4.3
5.9	8.0
1.7	2.0
3.7	4.8
3.1	6.8
1.7	9.1
3.2	1.5
32.5	52.9

Is there a significant average porosity difference between wells?

In this problem the comparison will be between two sample means, and this requires a slight rearrangement of the equations. The equation for t becomes

$$t = (\overline{X}_1 - \overline{X}_2)/\bar{s}\sqrt{1/N_1 + 1/N_2} \tag{2.4}$$

where

$\overline{X}_1$ and $\overline{X}_2$ are the sample means,
N_1 and N_2 are the number of observations in each set,
$\bar{s}$ is the pooled mean square estimate of σ,

$$\bar{s} = \{[\Sigma x_{i1}^2 - (\Sigma x_{i1})^2/N_1 + \Sigma x_{i2}^2 - (\Sigma x_{i2})^2/N_2]/(N_1 + N_2 - 2)\}^{1/2} \quad (2.5)$$

$$df = N_1 + N_2 - 2$$
$$= 10 + 10 - 2 = 18$$
$$\bar{s} = [(130.89 - 105.62 + 336.21 - 279.84)/18]^{1/2} = \pm 2.13$$
$$\overline{X}_1 = 32.5/10 = 3.25;\ \overline{X}_2 = 52.9/10 = 5.29$$
$$t = (5.29 - 3.25)/(2.13\sqrt{0.2}) = 2.14\ ^*$$

The calculated t is significant at the 95% level; therefore, the conclusion is that there is a population difference in porosity between the two wells, unless we have been victims of a 1-in-20 chance.

Tests of significance—the chi-square test

The t-test deals with measurements of continuous variables. There are other problems where the interest is merely in counting the number of cases that fall into specified categories. The information pertinent to this type of problem is called enumeration data because it is composed entirely of integers, and different procedures from those used with continuous variables are required. Chi-square (χ^2) is a distribution function which has been derived to handle enumeration data. It gives a test for determining whether an observed count differs significantly from an expected count.

The basis of the chi-square distribution is as follows: assume that a sample of "f" quantities, $q_1, q_2, q_3 \ldots, q_f$ is drawn at random from a normal population having zero mean and unit standard deviation. Then chi-square is the sum of squares of these "f" quantities:

$$\chi^2 = q_1^2 + q_2^2 + q_3^2 + \ldots + q_f^2$$

If we were to continue drawing samples from the same population, with each sample having f quantities, we could build up a histogram of χ^2. As the number of samples tends to infinity, the histogram tends to a smooth distribution function. This distribution function is called the χ^2 distribution. The form of the distribution depends upon sample size f, which determines the number of degrees of freedom.

Table A-2 shows the sampling distribution of chi-square. On examination of the table, you will see that the distribution depends upon degrees of freedom, just as in the case of the t-distribution.

Table A-2 shows values of χ^2 corresponding to various degrees of freedom and probability levels. The table is used much in the same way as the t-table. Thus at 7 degrees of freedom and a probability level of 95% the table gives $\chi^2 = 14.07$. This means that the chance that a sample χ^2 with 7 degrees of freedom will exceed 14.07 is just 1 in 20.

To test the significance of an observed χ^2 one follows much the same procedure that he would in making the t-test. A null hypothesis is set up and χ^2 is calculated from the sample data. This value is referred to the chi-square table at the appropriate degrees of freedom, and the probability is determined from the table. If the probability is less than 0.05, χ^2 is significant. If it is less than 0.01, χ^2 is highly significant.

To test for significance against the χ^2 distribution we need to calculate a sample χ^2, which we will refer to as a statistic, because it is calculated solely from a sample. Suppose there are k categories and that we have a sample of N observations and each observation must fall into one, and only one, category. After each observation has been placed in its proper category, we may count the frequencies in each category and denote the observed frequencies by f_1, f_2, f_3, etc. It follows that $\sum_{i=1}^{k} f_i = N$. In some situations there may be a theoretical frequency for each category which we will denote as F. Then $\sum_{i=1}^{k} F_i = N$. If we know the theoretical frequency, we may ask the question, "Do the observations disagree with the expected values?" The statistic χ^2 is defined as

$$\begin{aligned}\chi^2 &= \sum_{i=1}^{k} [(f_i - F_i)^2/F_i] \\ &= \sum \frac{(\text{observed value} - \text{expected value})^2}{\text{expected value}}\end{aligned} \tag{2.6}$$

The example problems to follow will show how this is applied.

Where the degrees of freedom are large – say greater than 30 – the expression $\mu = \sqrt{2\chi^2} - \sqrt{\text{df} - 1}$ can be applied. In this case μ is an approximation of t at infinite degrees of freedom.

The significance of μ is tested against the normal distribution which comprises the last line (df $= \infty$) of Table A-1. As an example assume $\chi^2 = 75$ and df $= 50$. Then $\mu = \sqrt{150} - \sqrt{99} = 2.30$. Interpolating in

Table A-1 we find that the probability corresponding to $\mu = 2.30$ is 0.0111; therefore, χ^2 is significant. When Table A-1 is used in this fashion, the probability value does not have a $\pm$ sign as in the two-sided t-test.

Two restrictions should be obeyed when using the χ^2 test. One is that the test should be used with caution if any of the expected or observed values is less than 5. When any of the categories has an expectancy less than 5, it usually can be grouped with another category or dropped from the table. This restriction does not have to be applied rigidly. When there are many degrees of freedom, it does not matter much if a few expectancies are less than five.

The other restriction is that an adjustment should be made when df = 1. This adjustment amounts to subtracting 0.5 from the term inside parentheses in the numerator of equation 2.6.

The examples following will illustrate uses of the χ^2 test:

Example 1

A drilling contractor's crews have the following record of manhours lost over a specific period of time because of absenteeism. There are equal numbers of men on each tour.

Tour	Daylight	Evening	Morning	Total
Manhours lost	61	72	93	226

Does this record indicate that the morning tour has a significantly higher rate of absenteeism than the other tours?

This is a case of equal expectancy because there are an equal number of men on each crew and the total hours worked by each crew is the same. The probability that a particular crew will lose a man-hour is 1/3, so the expected number of manhours lost per crew will be $1/3 \times 226 = 75.333$. The expected number is fictitious, of course, because the data are in integers and the expected value has a fractional remainder. If the calculations are set up in tabular form as in Table 2C the calculations will be facilitated.

The degrees of freedom, just as in the case of the t-test, depend upon the number of independent deviations. Although there are three devia-

TABLE 2C

Tour	Daylight	Evening	Morning	Total
Actual manhours lost	61	72	93	226.00
Expected manhours lost	75.333	75.333	75.333	226.00
Deviation	−14.333	−3.333	+17.667	0.00
χ^2 = (Deviation)^2 ÷ (Expected Value)	2.73	0.15	4.14*	7.02*

tions, the method of calculation implies that their total must be exactly zero. Hence, there are only two independent deviations, since when any two are known the third can be determined by difference. The number of degrees of freedom, then is $N - 1 = 2$.

If we enter the table with $\chi^2 = 7.02$ at df $= 2$ we will see that χ^2 is significant at the 5% probability level. This means that we can be 95% sure that there really is a difference between the records of the three tours. Table 2C is the most convenient form for χ^2 calculations because it shows the contribution of each category to χ^2. This provides the investigator the opportunity to get an approximate test of the significance of each category. Such a test is made by entering the table at df $= 1$ with the individual χ^2. This procedure shows that the deviations of the daylight tour and the evening tour are insignificant and the deviation of the morning tour is significant at the 5% level. This means that if the investigator decided that the morning tour had more than an average amount of absenteeism, he would have a 95% probability of being correct. Values of χ^2 which are significant at the 95% level are marked with one asterisk according to the convention mentioned in the previous section on t-tests.

Example 2

Many problems involve categories that do not have equal expectancies. Table 2D illustrates such a problem.

A drilling contractor has compiled a record of the accidents experienced by his employees over a specified period of time. He would like

TABLE 2D

Job classification	Average no. employed	No. of accidents	Rate per hundred employees	Expected number	Deviation	Contribution to χ^2
Floor men	75	30	40.00	32.143	−2.143	0.143
Derrick men	15	6	40.00	6.429	−0.429	0.029
Drillers	15	3	20.00	6.429	−3.428	1.828
Truck drivers and mechanics	14	12	85.71	5.999	+6.000	6.000 *
Total	119	51		51.000	0.000	8.000 *
General rate			42.857			

to know whether the accident frequency is greater or less than average in any particular job category.

There are four classifications for which deviations will be computed, so df = N − 1 = 3. Unequal expectancies are accounted for by prorating the accidents in each classification per each 100 employees. Thus, the accident rate per 100 rotary helpers is 30/75 × 100 = 40.00. The general rate per 100 employees = 51/119 × 100 = 42.857. Here, as in the previous example, this value is fictitious because obviously there cannot be a fraction of an accident.

The expected number of accidents in any classification is:

$$(\text{Observed number}/\text{Expected number per } 100) \times (\text{General rate per } 100) \qquad (2.7)$$

e.g. for floor men

$$\text{Expected number} = (30/40.000) \times 42.857 = 32.143$$

The deviation is the actual number minus the expected number:

$$\text{Deviation} = 30 - 32.143 = -2.143$$

The contribution of floor men to χ^2 is

$$(\text{Deviation})^2/(\text{Expected number}) = 0.143$$

The χ^2 contributions of each classification are computed in this manner and then summed. The total is checked for significance in the table.

Entering the table at df = 3 we see that $\chi^2 = 8.000$ is significant at the 5% level, so we conclude that there are significant differences between job classifications. The largest contribution to χ^2 is from truck drivers and mechanics. Checking the χ^2 contribution at df = 1, we see that the contribution of truck drivers and mechanics is significant at the 5% level. We might conclude, then, that the truck drivers and mechanics have a significantly higher than average accident frequency.

It is worth noting at this point that the most significant classifications are not necessarily those which show the greatest deviation. A classification may show a large deviation but fail to be significant because of the size of the numbers upon which it is based. As an example, the deviation for drillers is relatively high but it fails to reach significance when tested.

Tests of significance – the contingency test

The chi-square test applied to data which can be cross-classified is called a *contingency test.* It is simply a special application of the chi-square test; the same rules apply. Data for the contingency test are arranged in a *contingency table,* an example of which is shown in Table 2E.

Table 2E is the drilling record of four districts and it is desired to know how the districts compare. That is, are there differences between

TABLE 2E

Performance Category	District				Total
	A	B	C	D	
Wells drilled to TD on schedule	80	130	65	120	395
Wells drilled to TD 20% or more over schedule	12	35	17	13	77
Wells sidetracked to TD	8	16	2	7	33
Wells lost	20	32	5	35	92
Total	120	213	89	175	597

the districts, and are there any specific categories in which any of the districts excel or lag behind?

The table could be called a 4 × 4 contingency table because there are four rows and four columns. In this problem the rows refer to performance categories and the columns refer to districts.

Calculations proceed in the same way as in the other χ^2 tests. Expected frequencies are calculated from the product of the appropriate row and column totals divided by the overall total. Chi-square is then calculated and compared to the theoretical value for the proper degrees of freedom. The degrees of freedom, in general, are equal to one less than the number of rows times one less than the number of columns; so, letting r = no. of rows and c = no. of columns.

$$df = (r - 1) \times (c - 1) \tag{2.8}$$

When there is only one row the degrees of freedom will be c − 1.

The next step is to prepare a table of expected values. Expected values are computed from the following relation:

$$(\text{Expected value})_{r,c} = (\text{Row total} \times \text{Column total}) \div (\text{Grand total}) \tag{2.9}$$

For example, the expected value for wells that reached TD on schedule in District C (row 1, column 3) would be

$$(\text{Expected value})_{1,3} = (395)(89)/597 = 58.9$$

Table 2F shows the expected values for the example problem.

TABLE 2F

Performance Category	District				Total
	A	B	C	D	
Wells drilled to TD on schedule	79.4	140.9	58.9	115.8	395.0
Wells drilled to TD 20% or more over schedule	15.5	27.5	11.5	22.5	77.0
Wells sidetracked to TD	6.6	11.8	4.9	9.7	33.0
Wells lost	18.5	32.8	13.7	27.0	92.0
Total	120.0	213.0	89.0	175.0	597.0

TABLE 2G

Performance Category	District				Total
	A	B	C	D	
Wells drilled to TD on schedule	+0.6	−10.9	+6.1	+4.2	0.0
Wells drilled to TD 20% or more over schedule	−3.5	+7.5	+5.5	−9.5	0.0
Wells sidetracked to TD	+1.4	+4.2	−2.9	−2.7	0.0
Wells lost	+1.5	−0.8	−8.7	+8.0	0.0
Total	0.0	0.0	0.0	0.0	0.0

After the expected values have been tabulated, the deviations are calculated and tabulated in similar fashion as shown in Table 2G.

The deviation table provides a check on the arithmetic because all totals will be zero if the computations have been done correctly.

Equation 2.6 is used to calculate χ^2 as follows:

		Row	Column
$(0.6)^2/79.4$	= 0.004	(1)	(1)
$(10.9)^2/140.9$	= 0.843	(1)	(2)
$(6.1)^2/58.9$	= 0.632	(1)	(3)
$(4.2)^2/115.8$	= 0.152	(1)	(4)
$(3.5)^2/15.5$	= 0.790	(2)	(1)
$(7.5)^2/27.5$	= 2.045	(2)	(2)
$(5.5)^2/11.5$	= 2.630	(2)	(3)
$(9.5)^2/22.5$	= 4.011 *	(2)	(4)
$(1.4)^2/6.6$	= 0.297	(3)	(1)
$(4.2)^2/11.8$	= 1.495	(3)	(2)
$(2.9)^2/4.9$	= 1.716	(3)	(3)
$(2.7)^2/9.7$	= 0.752	(3)	(4)
$(1.5)^2/18.5$	= 0.122	(4)	(1)
$(0.8)^2/32.8$	= 0.020	(4)	(2)
$(8.7)^2/13.7$	= 5.525 *	(4)	(3)
$(8.0)^2/27.0$	= 2.370	(4)	(4)

$\Sigma\chi^2 = 23.404$ **

Entering the χ^2 table with nine degrees of freedom we see that $\chi^2 = 23.4$ is highly significant. That is, there is less than 1 chance in 100 that all the districts are performing the same. Checking the individual contributions we find that the row 2, column 4 and the row 4, column 3 contributions are significant at the 5% level. Consulting Table 2E we see that these correspond to "District D—wells drilled to TD 20% or more over schedule" and "District C—Wells lost," respectively. If we look at Table 2G it becomes apparent that there were significantly fewer wells in District D that ran over schedule and in District C there were significantly fewer wells lost.

Contingency tables can take many forms, and the computations are always made in the same way *except* for 2 × 2 tables. The 2 × 2 table has only one degree of freedom, so the restriction mentioned earlier in this discussion will apply. Errors are introduced if the χ^2 distribution is used when df = 1 because the data will always consist of whole numbers; consequently, the possible deviations will always be in steps of unity. The chi-square distribution is a continuous mathematical function and it should not be applied to discrete numbers, but statisticians have found that the χ^2 distribution can be used safely when df = 1 if a correction of one-half unit is applied to each deviation. The half unit is subtracted from the absolute value of the deviation so that Equation 2.6 becomes

$$\chi^2 = [\text{Abs (Observed Value} - \text{Expected Value)} - 0.5]^2 \div \text{Expected Value} \tag{2.10}$$

Contingency tables need not be restricted to a two-way classification. Triple or multiple classification tables are possible. Methods of computation would be the same as for two-way classifications.

OIL INDUSTRY REFERENCES

Perez-Rosales, Candelario: "A Simplified Method for Determining Specific Surface," Trans. AIME (1967), Vol. 240, p. I-1081.

Raynal, Jean C., Gstalder, Serge A., Sagot, Andre M. and Muckleroy, J. A.: "Organization of Field Tests and Evaluation of Tricone Bit Performance Using Statistical Analysis and Sonic Logs," *Journal of Petroleum Technology,* (April, 1971), p. 506.

3

Analysis of variance and multiple comparisons of means

Analysis of variance: general principles

Analysis of variance is a technique by which the variations associated with defined sources may be isolated and estimated. By "defined sources" is meant factors which we know or suspect to be contributing to the overall variation of an observed variable.

For example produced gas-oil ratio may vary with completion depth, well location and pressure drawdown, therefore these would be sources of variation, which we may wish to isolate and test for significance.

Whereas the t-test is used for testing whether the means of two sets of data are essentially equal, the analysis of variance is useful for testing for differences among the means of two or more populations. It also allows the sampling variation and the testing variation to be isolated and estimated. *The underlying principle of the analysis of variance is that if means of subgroups are greatly different, the variance of the combined groups is much larger than the variances of the separate groups.* The analysis of variance is, in fact, an analysis of differences in means.

Analysis of variance is a flexible technique that can be applied in many different ways depending on the situation. Some of the applications can become quite complex. The presentation given here represents only a brief resumé of the subject; therefore, in applying analysis of variance techniques to a particular problem the reader may find it necessary to turn to supplementary sources.

A significance test called the F-test is applied to pairs of variances. The statistic, F, is defined as

$$F = s_1^2/s_2^2 \tag{3.1}$$

where s_1^2 and s_2^2 are the two variances being compared. The procedure in applying an F-test is similar to that of the t-test; that is, a null hypothesis is set up and tested. Where the null hypothesis in the t-test is that two means are equal, the null hypothesis in the F-test is that two variances are equal.

The sampling distributions of F are shown in Tables A-3 and A-4 (Appendix). Table A-3 is the distribution of F at the 95% probability level and A-4 is the distribution at the 99% level. Values reported in the F table are the maximum values at that particular number of degrees of freedom and probability level; therefore any observed number that is larger than the tabulated value is significant at the particular probability level.

The F-test does not require that the degrees of freedom associated with the two variances be equal. Reference to Tables A-3 and A-4 will show there are degrees of freedom across the top of the table for the numerator and degrees of freedom in the left hand column for the denominator. For example, if

$$F = (s_1^2 \text{ @ df} = 3)/(s_2^2 \text{ @ df} = 9)$$

it would be tested against the table value in Column 3, row 9.

Analysis of variance makes use of a very important principle: when two or more independent sources of variation operate, their variances and degrees of freedom are additive. As an example, consider a set of porosities from core analysis. The set will have a certain amount of variance about a mean value. That variance will be caused by (a) sampling variance and (b) measurement error. Both of these sources of variance will be independent of one another and the total variance will simply be the sum of the two. So denoting sampling variance by σ_s^2 and error variance by σ_e^2 the total variance is

$$\sigma^2 = \sigma_s^2 + \sigma_e^2 \tag{3.2}$$

or, expressed as an estimate of variance from samples,

$$s^2 = s_s^2 + s_e^2 \tag{3.3}$$

Mention was made previously in the section on confidence limits that if a group of samples of size N has a variance of s^2, then the means of the samples will have a variance s^2/N. So, if N analyses are carried out on one sample and the results are averaged, the variance due to experimental error would be reduced to s_e^2/N and Equation 3.3 would become $s^2 = s_s^2 + s_e^2/N$. Likewise, if K samples are drawn from the population and N analyses are carried out on each, the equation would become $s^2 = (s_s^2 + s_e^2/N)/K = s_s^2/K + s_e^2/NK$. In practice, small samples will be used, and K and NK will be replaced by the appropriate degrees of freedom.

If there are K samples and N analyses on each sample, the variance within the jth sample caused by experimental error can be written $s_{ej}^2 = \sum_{i=1}^{N} (x_{ij} - \overline{X}_j)^2/(N-1)$ according to Equation 1.10. Such an expression is obtained for each sample and the variances are summed to get an estimate of the variance due to experimental error so that

Error Variance = Σ Sums of squares about sample means/Total degrees of freedom

$$= \sum_{i=1}^{N} \sum_{j=1}^{K} (x_{ij} - \overline{X}_j)^2/K(N-1) \qquad (3.4)$$

The numerator may be written more conveniently for computations as

$$\text{S.S. About Means} = \sum_{i=1}^{N} \sum_{j=1}^{K} x_{ij}^2 - \sum_{j=1}^{K} \left[\left(\sum_{i=1}^{N} x_{ij}\right)^2/N\right] \qquad (3.5)$$

The sum of squares about sample means usually is called the *sum of squares within samples.*

The sample variance will be estimated from the variance of the sample means. If the means of K samples are designated $\overline{X}_1, \overline{X}_2, \ldots, \overline{X}_K$ the variance between the sample means will be

$$\sum_{j=1}^{K} (\overline{X}_j - \overline{X})^2/(K-1) \qquad (3.6)$$

where $\overline{X}$ is the grand mean.

Equation 3.6 is an estimate of the sampling variance which in turn is an estimate of $\sigma_s^2 + \sigma_e^2/N$. This may be written:

$$N\sum_{j=1}^{K} (\overline{X}_j - \overline{X})^2/(K-1) \text{ is an estimate of } N\,\sigma_s^2 + \sigma_e^2 \qquad (3.7)$$

The expression $N \sum_{j=1}^{K} (\overline{X}_j - \overline{X})^2$ is called the *sum of squares between samples.* A more convenient form for computation is

$$\text{S.S. Between Samples} = \sum_{j=1}^{K} \left[\left(\sum_{i=1}^{N} x_{ij} \right)^2 / N \right] - \left(\sum_{j=1}^{K} \sum_{i=1}^{N} x_{ij} \right)^2 / NK \quad (3.8)$$

A third sum, the *total sum of squares,* will be required. It is the sum of squares of NK observations about the grand mean, $\overline{X}$. It has NK-1 degrees of freedom.

$$\text{Total sum of squares} = \sum_{j=1}^{K} \sum_{i=1}^{N} (x_{ij} - \overline{X})^2 \quad (3.9)$$

An expression more adaptable to computation is

$$\text{Total sum of squares} = \sum_{j=1}^{K} \sum_{i=1}^{N} x_{ij}^2 - \left(\sum_{j=1}^{K} \sum_{i=1}^{N} x_{ij} \right)^2 / NK \quad (3.10)$$

The first terms of Equations 3.5, 3.8 and 3.10 are referred to as the *crude sum of squares* and the second terms are called *the correction for the mean.*

The foregoing equations which comprise the elements of an analysis of variance may appear complex when written in mathematical notation, but the actual calculations are done in tabular form and, as the example problems later in this section will show, they are quite straightforward when done in that manner. At this point it will be useful to review the principles of variance analysis before proceeding to applications.

(1) Analysis of variance permits the isolation of sources of variance.

(2) The sources of variance may be tested for significance against the sampling distribution of F. The statistic, F, is defined as the ratio of the two variances being compared.

(3) When two or more *independent* sources of variance operate in a system of measurements, their variances and degrees of freedom are additive.

(4) In any system of measurements the total variance is the sampling variance plus the variance due to measurement error.

(5) If the variance of combined groups of samples is significantly larger than the variances of the separate groups, then the means of the subgroups are different.

(6) The sums of squares of three elements are required for an analysis of variance. These are (a) sum of squares within samples, (b) sum of squares between samples and (c) total sum of squares.

Utility of the F test in analysis of variance is based on the assumption that comparisons will be made between populations of the same variability but possibly of different means. The restriction that data sets must have similar variabilities within their individual measurements appears to be satisfied within practical limits by most comparative experimental data sets. The same is generally true for the requirement that the measurements be normally distributed. However, if the variances within the sets are widely different, a question arises as to whether a significant F ratio is due to differences in the means or differences in variances. For this reason a method is required to determine whether a series of variances from different samplings are homogeneous.

Bartlett's test for homogeneity of variances

The test of homogeneity of variances when K samples have been taken is based on the observation that the difference between K times the logarithm of the average variance and the sum of the logarithms of the individual variances is distributed as chi-square, χ^2, with K-1 degrees of freedom. (Refer to "Confidence Limits and Tests of Significance" for an explanation of the chi-square test). When equal numbers of measurements have been taken for each sample, this relationship is given by

$$\chi^2 = 2.3026\left[\mathrm{K}\log_{10}\bar{s}^2 - \sum_{j=1}^{K}\log_{10}s_j^2\right](\mathrm{N}-1) \qquad (3.11)$$

where

s_j^2, the group variance $= [\Sigma x_i^2 - (\Sigma x_i)^2/\mathrm{N}]/(\mathrm{N}-1)$
$\bar{s}^2$, the pooled variance $= \Sigma(\mathrm{N}-1)s_j^2/\Sigma(\mathrm{N}-1)$
N is the number of observations from each sample.
K is the number of samples.
degrees of freedom for $\chi^2 = \mathrm{K} - 1$

When unequal numbers of measurements, N_j, make up the different samples, χ^2 is computed from the relationship

$$\chi^2 = 2.3026\left[\sum_{j=1}^{K}(N_j - 1)\log_{10}\bar{s}^2 - \sum_{j=1}^{K}(N_j - 1)\log_{10}s_j^2\right] \quad (3.12)$$

The χ^2 values calculated by Equations 3.11 and 3.12 are slightly biased upward and may be corrected, if precision is critical, by dividing χ^2 by a factor,

$$C = 1 + (K + 1)/[3K(N - 1)] \quad (3.13)$$

if there are an equal number of measurements, or

$$C = 1 + 1/[3(K - 1)]\left[\sum_{j=1}^{K}1/(N_j - 1) - 1/\sum_{j=1}^{K}(N_j - 1)\right] \quad (3.14)$$

if there are an unequal number of measurements.

An application of the Bartlett test will be found in the section following.

There will be some situations where the primary interest will be in comparing means of the individual groups that comprise the analysis of variance. This special type of problem will be discussed later under "Comparison of More than Two Means."

Analysis of variance: one-way classification

The type of analysis just described constitutes the simplest form—the *one-way analysis of variance,* also known as a *hierarchic classification.* The hierarchic classification is illustrated schematically in Figure 3.1.

In the one-way analysis several samples are drawn from a popula-

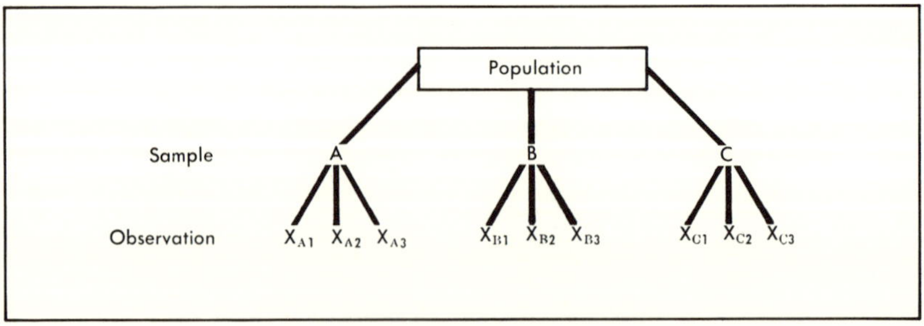

FIG. 3.1 Hierarchic classification.

tion. Each sample will be made up of independent observations of some variable x, as shown in Fig. 3.1. The purpose is to separate and evaluate sources of variation in the population. Table 3A summarizes the elements developed in the foregoing discussion.

TABLE 3A

Source of variation	Sum of squares	Degrees of freedom	Mean square	Mean square is estimate of
Between samples	$N \sum_{j=1}^{K} (\overline{X}_j - \overline{X})^2 = S_s$	$K - 1$	$S_s/(K - 1)$	$\sigma_e^2 + N\sigma_s^2$
Within samples	$\sum_{j=1}^{K} \sum_{i=1}^{N} (x_{ij} - \overline{X}_j)^2 = S_e$	$K(N - 1)$	$S_e/K(N - 1)$	σ_e^2
Total	$\sum_{j=1}^{K} \sum_{i=1}^{N} (x_{ij} - \overline{X})^2$	$NK - 1$		

The summations of Equations 3.5, 3.8, and 3.10 seem to be rather formidable when written with mathematical symbols, but they prove to be quite simple when the data are arranged tabularly as in Table 3B.

TABLE 3B

	Sample			
Observation	A	B	C	D
1	x_{A1}	x_{B1}	x_{C1}	x_{D1}
2	x_{A2}	x_{B2}	x_{C2}	x_{D2}
3	x_{A3}	x_{B3}	x_{C3}	x_{D3}
4	x_{A4}	x_{B4}	x_{C4}	x_{D4}
Total	Σx_A	Σx_B	Σx_C	Σx_D

Consider the following problem as an example.

Rock samples from a carbonate reservoir have been grouped into three classifications depending upon petrologic characteristics. Five core plugs have been selected randomly from each petrologic classification and the tortuosities measured by mercury injection according to the method of Burdine, et al. The question is: is the variation of tortuosity between rock types significantly different from the variation of tortuosity within rock types? A significant difference would help to justify the classifications physically as well as petrologically.

Table 3C shows the data arranged according to classification.

TABLE 3C

Observation	Tortuosity		
	Rock type A	Rock type B	Rock type C
1	4.9	2.9	5.1
2	3.5	2.3	5.3
3	3.1	2.9	2.9
4	4.5	2.7	5.3
5	2.9	2.9	4.7
Total:	18.9	13.7	23.3
Mean:	3.8	2.7	4.7

The elements required for calculation of the sums of squares are as follows:

$$C = \left(\sum_{j=1}^{K} \sum_{i=1}^{N} x_{ij}\right)^2 / NK = (\Sigma x)^2/\text{Total no. of observations}$$
$$= (18.9 + 13.7 + 23.3)^2/15 = 208.32$$

The constant, C, is the correction for the mean. It will be recognized as the second term of Equations 3.8 and 3.10.

The crude sum of squares between rock types is

$$\sum_{j=1}^{K} \left[\left(\sum_{i=1}^{N} x_{ij}\right)^2 / N\right] = \sum_{j=1}^{K} [(\text{Set sum})^2/(\text{No. in a set})]$$
$$= (18.9)^2/5 + (13.7)^2/5 + (23.3)^2/5 = 217.55$$

and the crude total sum of squares is

$$\sum_{j=1}^{K} \sum_{i=1}^{N} x_{ij}^2 = \Sigma x^2 = (4.9)^2 + (3.5)^2 + \ldots + (4.7)^2 = 225.03$$

The sum of squares between rock types, Equation 3.8

$$= \sum_{j=1}^{K} [(\text{Set sum})^2/(\text{No. in a set})] - C = 217.55 - 208.32$$
$$= 9.23$$

The sum of squares within rock types, Equation 3.5

$$= \Sigma x^2 - \sum_{j=1}^{K} [(\text{Set sum})^2/(\text{No. in a set})] = 225.03 - 217.55$$
$$= 7.48$$

The total sum of squares, Equation 3.10

$$= \Sigma x^2 - C = 225.03 - 208.32 = 16.71$$

For convenience the results usually are arranged as a table of variance so that numerical results can be presented as shown in Table 3D which is the table of variance for the example problem.

TABLE 3D

Source of variation	Degrees of freedom	Sum of squares	$s^2 = $ s.s./df	$F = s_1^2/s_2^2$
Between rock types	2	9.23	4.61	7.43**
Within rock types	12	7.48	0.62	1.00
Total	14	16.71	1.19	

Since the problem is to determine whether the variance of means between rock types is greater than the variance of means within rock types, the statistic to be tested will be $F = s_1^2/s_2^2 = 4.61/0.62 = 7.43$. The F-table (Table A-4) at two degrees of freedom in the numerator and twelve degrees of freedom in the denominator shows that F is significant at the 1% probability level. Therefore, it can be said with less than 1% probability of error that there really is a difference between rock types. By the same convention employed in the t-test, the double asterisk superscript in Table 3D denotes that F is highly significant.

It was stated before that the F-test is based on the assumption that variability is the same within each of the groups. This being true, the significance of F would be due entirely to differences in the means of the three rock types. Since there is a possibility, however, that the significance of F could be due to differences in variances, the Bartlett test should be made at this point in order to check for non-homogeneity of variances.

We would not expect the error variance to be any different for one rock type than for another, but the possibility should be considered that different technicians could have made the measurements and that one of these could have introduced a systematic measurement error. Another possibility is that a systematic error could have been introduced by the laboratory equipment, itself. An example would be a leak which occurs only at high pressure.

Chi-square for a group of variances is given by Equation 3.11 and the correction factor by Equation 3.13. The individual group variances are:

$$s_A^2 = [(4.9)^2 + (3.5)^2 + (3.1)^2 + (4.5)^2 + (2.9)^2 - (18.9)^2/5] \div 4 = 0.772$$
$$s_B^2 = [(2.9)^2 + (2.3)^2 + (2.9)^2 + (2.7)^2 + (2.9)^2 - (13.7)^2/5] \div 4 = 0.068$$
$$s_C^2 = [(5.1)^2 + (5.3)^2 + (2.9)^2 + (5.3)^2 + (4.7)^2 - (23.3)^2/5] \div 4 = 1.028$$

Table 3E summarizes the rest of the computations for the Bartlett test.

TABLE 3E

Rock type	N	s_j^2	$\log s_j^2$
A	5	0.772	−0.112
B	5	0.068	−1.168
C	5	1.028	0.012
		1.868	−1.268

$$C = 1 + (K+1)/[3K(N-1)] = 1 + 4/(3)(3)(4) = 1.111$$

$$\text{pooled variance, } \bar{s}^2 = \frac{\Sigma\,(N-1)s_j^2}{\Sigma\,(N-1)} = \frac{(4)(1.868)}{12} = 0.623$$

(which will be recognized as the "within groups" variance from Table 3D).

$$\log \bar{s}^2 = \log(.623) = -0.205$$
$$\chi^2 = 2.3026\,(N-1)\,[K \log \bar{s}^2 - \Sigma \log s_j^2]\,(1/C)$$
$$= (2.3026)(4)\,[(3)(-0.205) - (-1.268)]\,(1/1.111)$$
$$= 5.41$$

The value of χ^2 from Table A-2 at $K - 1 = 2$ degrees of freedom is 5.99 at the 95% confidence level, which means that the difference between rock types failed to reach significance at the 95% level; therefore it might be concluded that the variances are homogeneous and that the results of the F-test are valid. However, $\chi^2 = 5.41$ is so close to being significant that it would seem advisable to study larger samples before concluding the matter.

If χ^2 had been significant, this would have meant that the F-test had used an average of a group of variances which differed significantly among themselves. Whenever this occurs, the data should be regrouped into two or more groups, each with homogeneous variance, and the analysis of variance should be done on these new groupings.

A word of caution about use of the Bartlett test—it is very sensitive to departures from normality, so unless there is sufficient evidence to assume that the distributions of means are approximately normal, the results of the test should be accepted with reservation.

Other information is available from the analysis of variance. The *precision of measurement* can be estimated because the within-set variance is a pooled estimate of the variance of individual measurements. The estimated precision of measurement can be calculated by

precision $= t\sqrt{s_e^2}$, with t taken from the table at $1 - 1/2\alpha$ or $1/2\alpha$.

From the t-table at $K(N - 1) = 12$ degrees of freedom and 95% probability level, t is 2.18 so

$$\text{precision} = 2.18\sqrt{0.62} = \pm 1.72$$

Since the mean square between rock types is an estimate of $\sigma_e^2 + N\sigma_s^2$, the estimated standard deviation between rock types is

$$4.61 = 0.62 + 5\,s_s^2 \text{ and}$$
$$s_s = [(4.61 - 0.62)/5]^{1/2} = \pm 0.893$$

The standard deviation due to measurement error is

$$s_e = \sqrt{0.62} = \pm 0.787$$

Occasionally the investigator may want to know the confidence limits on the estimated variance. Statisticians have compiled tables which give approximations of these limits. If they are required they can be found in any book of statistical tables.

The explanation and example just given is for a *balanced design,* meaning that there are the same number of observations within each group; however, it is not always possible or desirable to use a balanced design. Whenever the design is unbalanced, some modifications have to be made. The sums of squares are calculated just as before, but one must be sure to divide the square of each group total by the number of observations in that group.

The mean square and F are calculated the same way as for a balanced design. The quantities estimated by the mean square, however, are a little different. In the balanced design the "between samples" (or "between groups") mean square is an estimate of $\sigma_e^2 + N\,\sigma_s^2$ and the "within samples" (or "within groups") mean square is an estimate of σ_e^2. If the experimental data are unbalanced, the quantities estimated by the mean squares are $\sigma_e^2 + \bar{n}\,\sigma_s^2$ and σ_e^2, respectively. The coefficient $\bar{n}$ is a function of the number of observations in the groups and is given by

$$\bar{n} = [(\text{Total no. of observations})^2 - \Sigma\, n_i^2]/[(\text{No. of groups} - 1) \times (\text{Total no. of observations})]$$

where n_i is the number of observations in the ith group. In most cases the average number of observations per group is a satisfactory approximation of $\bar{n}$.

There are two sources of variance in the above problem: variance between samples and variance within samples. Problems sometimes are encountered where there are three independent sources of variation. For example, in performing quality tests on some material such as cement or barites an investigator may suspect that there not only will be variation between bags and variation due to testing error, but also variation between shipments.

To test the hypothesis that there are three significant sources of variance, he may make Q tests on each of K bags selected randomly from each of N shipments, then arrange the results in a table of N rows and K columns with Q observations within each row and column. The three variances to be estimated will be σ_e^2 = residual or error variance; σ_c^2 = between-columns variance and σ_r^2 = between-rows variance. Equations for the sums of squares will be:

$$\text{Between-rows S.S.} = QK \sum_{i=1}^{N} (\overline{X}_i - \overline{X})^2 = \sum_{i=1}^{N} S_i^2/QK - S^2/QNK \qquad (3.15)$$

where

S_i = sum of observations in the ith row

S = total sum of observations

$$\text{Total S.S. between columns and rows} = Q \sum_{i=1}^{N}\sum_{j=1}^{K} (\overline{X}_{ij} - \overline{X})^2 = \sum_{i=1}^{N}\sum_{j=1}^{K} S_{ij}^2/Q - S^2/QNK \qquad (3.16)$$

where

S_{ij} = the sum of observations in the ith row and j[th] column.

$$\text{Total S.S.} = \sum_{i=1}^{N}\sum_{j=1}^{K}\sum_{t=1}^{Q} (x_{ijt} - \overline{X})^2$$

$$= \sum_{i=1}^{N}\sum_{j=1}^{K}\sum_{t=1}^{Q} x_{ijt}^2 - S^2/QNK \qquad (3.17)$$

The degrees of freedom for row sum of squares is N − 1. Degrees of freedom for sum of squares within columns and rows is N(K − 1), so the degrees of freedom for the total sum of squares between rows and columns is (N − 1) + N(K − 1) = NK − 1. The degrees of freedom on overall sum of squares is QNK − 1, consequently the degrees of freedom on the residual sum of squares is (QNK − 1) − (NK − 1) = NK(Q − 1).

Table 3F summarizes the elements of the one-way classification with three sources of variance.

TABLE 3F

Source of variation	Sum of squares	Degrees of freedom	Mean square = s.s./df	Quantity estimated by mean square
(1) Between rows	Eq. (3.15)	N − 1		$\sigma_e^2 + Q\sigma_c^2 + QK\sigma_r^2$
(2) Between columns within rows	(3) − (1)	N(K − 1)		$\sigma_e^2 + Q\sigma_c^2$
(3) Total between columns	Eq. (3.16)	NK − 1		
(4) Residual	(5) − (3)	NK(Q − 1)		σ_e^2
(5) Total	Eq. (3.17)	NKQ − 1		

The quantity estimated by mean square shown in Table 3F refers to a balanced design (i.e. an equal number of observations in each group and subgroup). If there are unequal numbers, the "between rows" mean square is an estimate of $\sigma_e^2 + \bar{n}_2\sigma_c^2 + \bar{n}_3\sigma_r^2$; the "between columns" mean square is an estimate of $\sigma_e^2 + \bar{n}_1\sigma_c^2$ and the residual mean square is an estimate of σ_e^2. Exact equations for $\bar{n}_1$, $\bar{n}_2$ and $\bar{n}_3$ are given in those statistical texts which have a comprehensive treatment of analysis of variance. The coefficients $\bar{n}_1$ and $\bar{n}_2$ can be approximated by the mean number of observations per column and $\bar{n}_3$ by the mean number of ob-observations per row.

An unbalanced design may be more desirable in some situations because the number of tests required to maintain a balanced design may become prohibitive. Following is a simple problem involving tests on oil-well cement which will illustrate the unbalanced design.

Three shipments of cement were tested for consistency of quality. The test carried out was the ASTM test of tensile strength. In this test tensile strength is measured on three standard briquettes of neat cement which have been cured by a standard procedure for 7 days. The test

TABLE 3G

	Bag		Total
Shipment	1	2	
1	675	654 735	2,064
2	760	788 796	2,344
3	915	796 817	2,528
			6,936

value is the average of the three measurements in pounds per square inch.

Two bags were selected at random from each of the three shipments. One bag was tested twice and the other bag was tested only once, hence the unbalanced design. Table 3G is a tabulation of the test results. In Table 3G there are three rows representing shipments and two columns representing bags (or samples). Within column 1 there is one test result and within column 2 there are two. There are $N - 1 = 2$ degrees of freedom for shipments and $N(K - 1) = 3$ degrees of freedom for samples. The number of tests, Q, are unequal, there being one test of bag 1 and two tests of bag 2 from each shipment. We will use $\overline{Q} = (1 + 2) \div 2 = 1.5$, so the degrees of freedom for residual variance will be $NK(\overline{Q} - 1) = (3)(2)(1.5 - 1) = 3$.

By Equation 3.15 the between-shipments sum of squares is

$$[(2{,}064)^2 + (2{,}344)^2 + (2{,}528)^2] \,/\, (1.5)(2) - (6{,}936)^2 \,/\, (1.5)(3)(2) = 36{,}395.$$

The total sum of squares between bags (samples) is given by Equation 3.16 as

$$(675)^2 + (760)^2 + (915)^2 + [(654 + 735)^2 + (788 + 796)^2 + (796 + 817)^2]/2 - (6{,}936)^2 \,/\, (1.5)(3)(2) = 45{,}179$$

The total sum of squares according to Equation 3.17 is

$$(675)^2 + (760)^2 + (915)^2 + (654)^2 + (735)^2 + (788)^2 + (796)^2 + (796)^2 + (817)^2 - (6{,}936)^2 \,/\, (1.5)(3)(2) = 48{,}712$$

TABLE 3H

Source of variation	Sum of squares	Degrees of freedom	Mean square	F
Shipments	36,395	2	18,198	15.45**
Samples	8,784	3	2,928	2.49
Residual	3,533	3	1,178	
Total	48,712	8		

The sum of squares between bags within shipments is 45,179 − 36,395 = 8,784, and the residual sum of squares is 48,712 − 45,179 = 3,533.

The table of variance is shown as Table 3H. In the F-test, the variance between shipments is significant at the 99% confidence level whereas the variance between samples within shipments fails to reach significance. We may conclude, therefore, that the physical properties of the cement really are different for the different shipments.

The quantities estimated by the mean squares are as follows,

$$\text{Residual MS} = s_e^2$$
$$s_e = \sqrt{1{,}178} = \pm 34.3 \text{ psi}$$
$$\text{Samples MS} = s_e^2 + 1.5\ s_b^2$$
$$s_b = [(2{,}928 - 1{,}178)\ /1.5]^{1/2} = \pm 34.2 \text{ psi}$$
$$\text{Shipments MS} = s_e^2 + 1.5\ s_b^2 + (1.5)(2)s_s^2$$
$$s_s = \{[18{,}198 - 1{,}178 - (1.5)(1{,}167)] \,/\, (1.5)(2)\}^{1/2} = \pm 71.3 \text{ psi}$$

where s_e is the standard error of the estimate, s_b is the standard deviation of the samples (i.e. bags) and s_s is the standard deviation of the shipments.

The analysis for a larger number of sources of variation can be done by following the same principles employed in the method for three sources of variation. For these more complex situations the engineer should refer to a statistical textbook.

Analysis of Variance: Two-Way Cross-Classification

The previous section considered a one-way or hierarchic classification. Another type of problem involves a *two-way classification*. The two-way problem is the simplest application of a technique called *cross-classification*. Cross-classification is useful when each observation in a set of data can be classified independently with respect to all sources of variation.

In a two-way problem the variation attributable to two distinct categories is investigated. For example suppose there are four technicians working in a laboratory and we want them to try four different methods for testing drilling mud viscosity. The aim is to determine whether there is any significant difference in results turned out by the four test methods and at the same time to find out whether the four lab technicians are performing equally well. The independent sources of variation would be analysts and test method.

Obviously, we would want to make sure that each technician used the same bulk mud at the same temperature for each of the four tests, because we wouldn't want sample variation to get into the analysis. It would be advisable, however, to use several samples of the same mud, but labeled differently, so that the technicians would not know they were measuring the same mud. This would help to eliminate biased results. The results of this type of test could be arranged in a two-way table as shown in Table 3I.

As a numerical example of a two-way problem, consider three temperature-compensated pressure bombs that are calibrated at controlled pressures and temperatures. Results of the calibrations are shown in Table 3J. The question is: Do all of the instruments have equal accuracy and do any of them show a significant tendency to drift with temperature?

TABLE 3I

Test method	Analyst A	B	C	D	Total
1	x_{A1}	x_{B1}	x_{C1}	x_{D1}	Σx_1
2	x_{A2}	x_{B2}	x_{C2}	x_{D2}	Σx_2
3	x_{A3}	x_{B3}	x_{C3}	x_{D3}	Σx_3
4	x_{A4}	x_{B4}	x_{C4}	x_{D4}	Σx_4
Total	Σx_A	Σx_B	Σx_C	Σx_D	Σx_{ij}

TABLE 3J

Temp., °F.	Instrument A	Instrument B	Instrument C	Total
100	2,109	2,113	2,106	6,328
200	2,115	2,116	2,105	6,336
300	2,122	2,117	2,108	6,347
Total	6,346	6,346	6,319	19,011

Since the calibration pressure in all cases is 2,100 psi the calculations can be simplified by subtracting 2,100 from each number.

Table 3J is arranged so that rows represent temperature and columns represent instruments, so when reference is made to rows and columns they will be considered in this context.

If there are N rows and K columns the equations for sums of squares will be:

S.S. between rows =

$$\Sigma(\text{Row totals})^2/(\text{No. in a row}) - (\Sigma x)^2/(\text{Total no. of observations}) \quad (3.18)$$

S.S. between columns =

$$\Sigma(\text{Column totals})^2/(\text{No. in a column}) - (\Sigma x)^2/(\text{Total no. of observations}) \quad (3.19)$$

Total S.S. =

$$\Sigma x^2 - (\Sigma x)^2/(\text{Total no. of observations}) \quad (3.20)$$

The elements of a two-way analysis of variance are tabulated in Table 3K.

TABLE 3K

Source of variation	Sum of squares	Degrees of freedom	Mean square = s.s./df	Quantity estimated by mean square
(1) Between rows	Eq. (3.18)	N − 1		$\sigma_e^2 + K\sigma_r^2$
(2) Between columns	Eq. (3.19)	K − 1		$\sigma_e^2 + N\sigma_c^2$
(3) Residual	(4) − (2) − (1)	(N − 1)(K − 1)		σ_e^2
(4) Total	Eq. (3.20)	NK − 1		

Applying the data from Table 3J and subtracting 2,100 from each value as specified earlier:

$$C = (\Sigma x)^2/(\text{Total no. of measurements})$$
$$= [19{,}011 - (9)(2{,}100)]^2/9 = 1369$$
$$\text{Total sum of squares} = [(9)^2 + (15)^2 + \ldots + (8)^2] - 1{,}369$$
$$= 260$$
$$\text{Sum of squares between temperatures}$$
$$= (1/3)[(28)^2 + (36)^2 + (47)^2] - 1{,}369 = 61$$
$$\text{Sum of squares between instruments}$$
$$= (1/3)[(46)^2 + (46)^2 + (19)^2] - 1{,}369 = 162$$
$$\text{Residual sum of squares}$$
$$= 260 - 61 - 162 = 37$$

The residual sum of squares includes all variance, from whatever sources, remaining after the column and row effects have been removed. If the column and row variances are completely independent, the residual mean square will be an estimate of the overall experimental error.

Table 3L is an analysis of variance table for the example problem.

TABLE 3L

Source of variation	Sum of squares	df	s^2 Mean square	$F = s^2/s_e^2$
Between temperatures	61	2	30.5	3.32
Between instruments	162	2	81.0	8.80*
Residual	37	4	9.2	1.00
Total	260	8	32.5	

The F-test will be used to test the significance of row and column variances. In making the test one would compare the "between temperatures" and "between instruments" F's of Table 3L to $F_{.95}$ (2,4) = 6.94 and $F_{.99}$ (2,4) = 18.0 from Tables A-3 and A-4, (Appendix) respectively.

When the F's are checked against these values, it is found that the between-temperatures variance is not significant, but the between-instruments variance is significant at the 95% level. Thus, one may say that the temperature drift of the three instruments is not significant but there is a significant difference in the pressure readings of the three. From the data it would appear that instrument C reads with less error than instruments A and B.

The overall precision of measurement is $t_{1/2a}\sqrt{s_e^2} = 2.31\sqrt{9.2} = \pm 7.0$ psi at the 95% confidence level.

Standard deviations attributable to the individual sources are:

Residual error: $s_e = \sqrt{9.2} = \pm 3.03$ psi

Between instruments: $s_I = [(81.0 - s_e^2)/N]^{1/2} = [(81.0 - 9.2)/3]^{1/2} = \pm 4.89$ psi

Between temperatures: $s_T = [(30.5 - s_e^2)/N]^{1/2} = [(30.5 - 9.2)/3]^{1/2} = \pm 2.66$ psi

Analysis of Variance: Interaction Effect

In a set of experiments with two defined sources of variation the effect of the variables may not be perfectly independent. If they are not, there is said to be an *interaction effect*. In other words, interaction is an effect in combination which is different from the sum of the separate effects. Interaction keeps the sums of squares from being truly additive. In some problems there will be a question as to whether the two sources of variation really are independent, and it will be desirable to measure the interaction effect and test its significance.

"Interaction" is the same as the residual sum of squares in the two-way analysis of variance but it is not called residual because it will consist of the combined effects of interaction and measurement error. To separate the two, it will be necessary to make a further estimate of variance, "within groups" to account for the measurement error. The remainder, after the effect of measurement error is removed, is the interaction effect. It accounts for those differences among means which cannot be accounted for by constant shifts in row means and column means.

As an example, consider another problem on pressure bomb calibration. Suppose it is required to know whether a bomb's accuracy is affected by temperature and magnitude of pressure being measured and whether the pressure and temperature effects are independent. An experiment would be designed for a cross-classification between temperature and pressure with provision for testing the interaction effect.

This experiment would provide several measurements of pressure at several controlled temperatures and pressures. The measurements should be made randomly to ensure that they are completely independent. The results of an experiment of this type are shown in Table 3M. The calibration pressure has been subtracted from the measured pressure so that the numbers shown in 3M are the differences between actual and measured pressures.

TABLE 3M

		Calibration pressure, psi			
		2,100	2,600	3,100	Total
Temperature, °F.	100	9, 12, 10 = 31	3, 6, 5 = 14	−2, 0, 1 = −1	44
	200	15, 16, 14 = 45	10, 12, 12 = 34	6, 5, 8 = 19	98
	300	22, 24, 21 = 67	16, 16, 18 = 50	10, 11, 12 = 33	150
Total		143	98	51	292

The repeat measurements at each pressure and temperature are called *replicates*. Replicates are necessary because they provide an estimate of measurement error that is independent of interaction effect. If there are K columns, N rows and Q replicates the equations for sums of squares of rows and columns will be the same as for the two-way classification, but there will be an additional sum of squares which will be

$$\text{S.S. of replicates} = \Sigma(S_t)^2/Q - (\Sigma x)^2/QNK \qquad (3.21)$$

where S_t is the subtotal of a set of replicates.

The generalized form for the interaction table is as shown in Table 3N.

TABLE 3N

Source of variation	Sum of squares	Degrees of freedom	Mean square = s.s./df	Mean square is an estimate of
(1) Between rows	Eq. (3.18)	$N-1$		$\sigma_e^2 + KQ\sigma_r^2$
(2) Between columns	Eq. (3.19)	$K-1$		$\sigma_e^2 + NQ\sigma_c^2$
(3) Interaction	(4)-(2)-(1)	$(N-1)(K-1)$		$\sigma_e^2 + Q\sigma_I^2$
(4) Subtotal	Eq. (3.21)	$NK-1$		
(5) Residual	(6)-(4)	$NK(Q-1)$		σ_e^2
(6) Total	Eq. (3.29)	$NKQ-1$		

For the example problem,

Total sum of squares $= (9)^2 + (12)^2 + \ldots + (12)^2 - (292)^2/27 = 1{,}134.07$

Sum of squares of replicates $= (1/3)[(31)^2 + (45)^2 + \ldots + (33)^2] - (292)^2/27 = 1{,}101.40$

Residual sum of squares = measurement error $= 1{,}134.07 - 1{,}101.40 = 32.67$

Sum of squares between temperatures $= (1/9)[(44)^2 + (98)^2 + (150)^2] - (292)^2/27 = 624.29$

Sum of squares between pressures $= (1/9)[(143)^2 + (98)^2 + (51)^2] - (292)^2/27 = 470.29$

Sum of squares for interaction $= 1{,}101.40 - 470.29 - 624.29 = 6.82$

Results of the computations are tabulated in Table 3O.

The significance of computed F's for "between temperatures" and "between pressures" variances will be tested against the tabled values

TABLE 3O

Source of variation	Sum of squares	Degrees of freedom	Mean square	F
Between temperatures	624.29	2	312.14	171.50**
Between pressures	470.29	2	235.14	129.20**
Interaction	6.82	4	1.70	0.93
Subtotal	1,101.40	8		
Experimental error	32.67	18	1.82	1.00
Total	1,134.07	26		

(Tables A-3 and A-4), $F_{.95}$ (2,18) = 3.55 and $F_{.99}$ (2,18) = 6.01. The computed F for "interaction effect" will be compared to $F_{.95}$ (4,18) = 2.93 and $F_{.99}$ (4,18) = 4.58.

These comparisons show that there is a highly significant variance between temperatures and between pressures but there is no significant interaction effect; thus the temperature and pressure effects are completely independent.

Whenever there is no interaction effect, as in the present problem, the "interaction" mean square and the "residual" mean square both are unbiased estimates of the variance due to measurement error. This indicates that pooling of the two may improve the estimate of σ_e^2. A rule of thumb is that pooling will be useful whenever the ratio s_I^2/s_e^2 is less than twice the 50th percentile value of F from the tables. (Tables of percentiles of the F distribution are available from most books of statistical tables).

To pool s_I^2 and s_e^2, add the sums of squares of "interaction" to the sums of squares of "residual" to get the new residual sum of squares. The new residual mean square is (new residual s.s.)/[(df for interaction) + (df for residual)]. The new residual mean square is then used to recalculate F for the column and row effects. In the present problem s_I^2 is close enough to s_e^2 that pooling would have no appreciable effect upon the estimate of s_e^2. If pooling was used, the degrees of freedom associated with the new estimate of s_e^2 would be 22, therefore the values $F_{.95}$ (2,22) = 3.44 and $F_{.99}$ (2,22) = 5.72 would be used to make comparisons in the F-test.

The mean squares from Table 3O provide the following standard deviation estimates:

Measurement error: $s_e = \sqrt{1.82} = \pm 1.35$ psi

Between temperatures: $s_T = [(312.14 - 1.82)/(3)(3)]^{1/2} = \pm 5.87$ psi

Between pressures: $s_P = [(235.14 - 1.82)/(3)(3)]^{1/2} = \pm 5.09$ psi

The 95% confidence limits on the precision of measurement are $2.10\sqrt{1.82} = \pm 2.83$ psi.

Analysis of variance – cross-classification with disproportionate observations

In the interaction problem just described there were an equal number of replicates in each cell. When a cross-classification problem arises where there is an unequal number of observations in the cells, rigorous analysis becomes difficult. The addition theorem for sums of squares does not hold in this situation. Occasionally this will be demonstrated forcefully when the sums of squares due to the independent sources of variation will add to more than the total sum of squares. The rigorous methods for handling disproportionate numbers of observations are beyond the scope of this discussion; however, when the disproportion is not large, a good approximate analysis can be obtained by the following procedure.

(1) Compute the arithmetic average of the observations in each cell.
(2) Do the analysis of variance using the cell averages. Include interaction if desired, but do not compute the residual mean square.
(3) Return to the original data and use a one-way analysis to compute the "within cells" mean square.
(4) Multiply the "within cells" mean square by the reciprocal harmonic mean number of observations and use this as the residual mean square for F-tests.

For example refer to the data of Table 3P. These data are permeabilities from two wells arranged into two depth intervals. The question is whether there is a difference between wells and between depths.

TABLE 3P

Depth interval	Well	
	A	B
I	20; 35; 68; 125; 30	15; 12; 32; 18
II	720; 82; 900	1000; 95; 630; 170; 300

TABLE 3Q

Depth Interval	Well A	Well B	Total
I	55.6 (5)	19.3 (4)	74.9
II	567.3 (3)	439.0 (5)	1,006.3
Total	622.9	458.3	1,081.2

Table 3Q is the same data reduced to cell means. The number in parentheses in each zone is the number of observations represented by the mean.

Total sum of squares = $\Sigma x^2 - (\Sigma x)^2/N$
$= (55.6)^2 + (19.3)^2 + (567.3)^2 + (439.0)^2 - (1081.2)^2/4$
$= 225{,}766$

Sum of squares between wells $= (1/2)[622.9)^2 + (458.3)^2] - (1{,}081.2)^2/4$
$= 6{,}773$

Sum of squares between depths $= (1/2)[(74.9)^2 + (1{,}006.3)^2] - (1{,}081.2)^2/4$
$= 216{,}876$

Interaction sum of squares $= 225{,}766 - 6{,}773 - 216{,}876$
$= 2{,}117$

Total sum of squares (original data) $= \Sigma x^2 - (\Sigma x)^2/$(total no. of observations)
$= (20)^2 + (35)^2 + \cdots + (300)^2 - (4{,}252)^2/17$
$= 1{,}820{,}940$

Sum of squares between cells (original data) = Σ [(cell total)^2/(no. of observations in cell)] − $(\Sigma x)^2$/(total no. of observations)
$= (278)^2/5 + (77)^2/4 + (1{,}702)^2/3 + (2{,}195)^2/5 - (4{,}252)^2/17$
$= 882{,}645$

Sum of squares within cells (original data) $= 1{,}820{,}940 - 882{,}645$
$= 938{,}295$

The degrees of freedom associated with the "within cells" sum of squares are

(Total No. of observations − 1) − [(No. of rows) (No. of columns) − 1] = (17 − 1) − [(2) (2) − 1] = 13, therefore the mean square is 938,295/13 = 72,177. The reciprocal of the harmonic mean is

$$[1/(\text{No. of cells})]\ (1/n_1 + 1/n_2 + \cdots + 1/n_n) = 1/4\ (1/5 + 1/4 + 1/3 + 1/5) = 0.245$$

The adjusted mean square to be used in F-tests is (0.245) (72,177) = 17,683.

The complete analysis of variance results are shown in Table 3R. The results show a highly significant difference between depth intervals but no significant difference between wells.

TABLE 3R

Source of variation	Sum of squares	Degrees of freedom	Mean square	F
Between wells	6,773	1	6,773	0.38
Between depths	216,876	1	216,876	12.26**
Interaction	2,117	1	2,117	0.12
Total	225,766	3		
Residual		13	17,683	

Analysis of variance – models I and II

Each of the examples of analysis of variance in the previous sections is *Model I.* There is another type, *Model II,* which is preferable to Model I in certain situations. Model II will not be demonstrated here, because it seems unlikely that it would be required in most petroleum engineering applications.

In Model I the groups making up the total sample are particular groups in which we are interested. We want to test these groups to see whether they are significantly different. On the other hand, in Model II we would not be especially interested in the groups that happen to be in the sample; our concern would be to estimate the variance of the infinite population of group means from which our sample was drawn.

The difference lies, then, in our purpose; and this purpose will dictate the method of sampling. When the experimenter knows at the outset what groups are to be included in his sample, he will draw random samples from each of those groups and will get a Model I analysis of variance. When there is a population composed of n groups and the experimenter wants to study any four of these groups, he will select any four at random then draw replicate samples from within each of those four. He will then have the elements for a Model II analysis of variance.

Either model can be used in any of the one-way or heirarchic classifications described in prior sections. In fact, mixed models are common in some types of statistical problems. Partitioning of the sums of squares and other computations are identical. The only difference comes in making the F-test.

It has been seen that with Model I the test is always between the residual variance and the variance due to some other category of classification. With Model II the test does not always follow this pattern. The engineer who is interested in Model II applications should refer to one of the newer texts on analysis of variance.

Comparison of more than two means

Analysis of variance tells us whether there is a significant difference between means when several means are grouped together, and Bartlett's test tells whether the variances of these means are homogeneous. The combination of these two methods, then, permits us to decide whether the means represent the same population or several different populations, but it does not tell us whether *all* the means are different from each other.

It would be possible to investigate the differences among, say 5 means by performing a t-test on each of 10 pairs of means, but there are two reasons why this is not good practice. First, each separate comparison that is made contributes to a cumulative decrease of the confidence level so that successive t-tests made at the 95% confidence level would be misleading. Second, there will be much loss of precision in estimating the variance if the measurements from only the two groups being compared are used.

A much more reliable method is to test the hypothesis of equal means by comparing the range of the means to the within-groups sum of squares. This can be done with another test statistic,

$$q = w/(\bar{s}/\sqrt{N}) \tag{3.22}$$

where w is the range of a group of means or the difference between two means that are being compared; $\bar{s}^2$ is the pooled or "within groups" mean square and N is the size of each sample. The denominator $\bar{s}/\sqrt{N}$ is an estimate of $\sigma/\sqrt{n}$ and is the standard deviation of $\overline{X}$'s for samples of size N drawn from the same population. Distributions of q at several probability levels are shown in Table A-5. You will note that the heading in Table A-5 does not include the divisor $\sqrt{N}$. This is because the table represents the distribution of the ratio for k single observations (N = 1) from a normal population. The statistic must be modified by the factor $\sqrt{N}$ to apply it to the means of N observations each.

The q-value is applied in a manner similar to the t statistic. It can be used to test the significance of a difference between means, $\bar{X}_a - \bar{X}_b$, or it can be used to determine confidence limits for that difference. Confidence limits are the most descriptive, and they can be tabulated with ease. All one must remember is that the difference, $\bar{X}_a - \bar{X}_b$, is not significant if the confidence limits include zero. This is sometimes referred to as Tukey's test after J. W. Tukey, the originator.

Comparisons also can be made between single means and grouped means, such as $[(\bar{X}_1 + \bar{X}_2)/2] - \bar{X}_3$, etc.

It will be noted that Table A-5 has υ rows and K columns, representing the degrees of freedom associated with $\bar{s}$ and the number of means, respectively. The degrees of freedom associated with $\bar{s}$ are given by K(N − 1).

In the problem on rock properties (Table 3C) it was shown that there are significant differences among the three means but it was not shown that every mean is significantly different from every other mean. This question can be answered by comparison of means.

There are three means, each made up of five observations and the degrees of freedom associated with $\bar{s}$ are 12, so K = 3, $\nu = 12$ and N = 5. Referring to Table A-5 we find that $q_\alpha = 3.77$, where $\alpha = .05$. This means that the expected value of q at the 95% confidence level is 3.77. The 95% confidence limits on any comparison then are

$$w = q_\alpha \bar{s}/\sqrt{N} = (3.77)(\sqrt{0.62})/\sqrt{5} = 1.33$$

$$\text{then } \bar{X}_A - \bar{X}_B = 3.8 - 2.7 = 1.1 \text{ and the 95\% confidence limits}$$

$$\text{on } \bar{X}_A - \bar{X}_B \text{ are } 1.1 \pm 1.33 = -0.23 \text{ and } 2.43.$$

These limits include zero, therefore $\bar{X}_A$ and $\bar{X}_B$ are not significantly different at the 95% probability level. Table 3S lists all the comparisons that could be made in that problem.

These results show that, at the 95% confidence level, there is no difference between rock types A and C and that they may as well be grouped together.

It was stated earlier that use of the t-test is not recommended for

TABLE 3S

Population comparison	Confidence limits
$\overline{X}_A - \overline{X}_B$	−0.23 and 2.43
$\overline{X}_A - \overline{X}_C$	−2.23 and 0.43
$\overline{X}_B - \overline{X}_C$	−0.67 and −3.33
$[(\overline{X}_A + \overline{X}_C)/2] - \overline{X}_B$	0.22 and 2.88
$[(\overline{X}_A + \overline{X}_B)/2] - \overline{X}_C$	−2.78 and −0.12
$[(\overline{X}_B + \overline{X}_C)/2] - \overline{X}_A$	−1.43 and 1.23

multiple comparisons. There is no reason, however, that it cannot be used for one particular comparison, provided it is used for no other. The comparison to be made should be specified beforehand.

The confidence limits on the single comparison will be $(\overline{X}_a - \overline{X}_b) \pm \bar{s}\ t_{1/2\alpha} \sqrt{\Sigma a^2/N}$, where Σa^2 is the summation of squares of the coefficients a_1, a_2, etc. corresponding to $\overline{X}_1$, $\overline{X}_2$, etc. That is, for $\overline{X}_a - \overline{X}_b$, $\Sigma a^2 = (1)^2 + (-1)^2 = 2$, and for $[(\overline{X}_a + \overline{X}_b)/2] - \overline{X}_C$, $\Sigma a^2 = (1/2)^2 + (1/2)^2 + (-1)^2 = 1.5$. To illustrate, the statistic t will be used in comparing $\overline{X}_A$ and $\overline{X}_C$ from the example problem. At df = 12 and 95% probability level, t = 2.18. The difference, $\overline{X}_C - \overline{X}_A = 4.7 - 3.8 = 0.9$, and the 95% confidence limits are $0.9 \pm (0.79)(2.18)\sqrt{2/5} = -0.19$ and 1.99. The difference between $\overline{X}_A$ and $\overline{X}_C$ is found to be insignificant, which confirms the conclusion of the q test.

Relation of statistical tests

As the engineer works with the various tests of significance, he will probably notice similarities between the different tests. The F, t and χ^2 tests are, in fact, interrelated. Chi-squared is related to the bottom row of the F table where the number of degrees of freedom for the denominator variance is infinity. Along this row F is χ^2 divided by the number of degrees of freedom. Thus, from the F table, for P = 0.01, $df_1 = 8$, $df_2 = \infty$, F is 2.51, or the result of dividing 20.1 by 8.

Where there is only one degree of freedom, χ^2 is simply the square of t at infinite degrees of freedom, e.g.,

$$\chi^2 = 1.96^2 = 3.84 \text{ at } P = 0.05$$

Likewise, $t^2 = F$ when one degree of freedom is associated with the numerator.

An interesting characteristic of χ^2 is that the sum of a χ^2 with df_1 degrees of freedom and a χ^2 with df_2 degrees of freedom is a χ^2 with $df_1 + df_2$ degrees of freedom. Also, df is the average or expected value of a χ^2 with df degrees of freedom.

OIL INDUSTRY REFERENCES

Testerman, J. D.: "A Statistical Reservoir Zonation Technique," Trans., AIME (1962) p. 889.

Warren, J. E. and Price, H. S.: "Flow in Heterogeneous Porous Media," SPE Journal (September, 1961) Vol. 1, p. 153.

4

Linear relationships between variables

Much of the data analysis faced by petroleum engineers concerns one variable whose magnitude is dependent upon the magnitude of some other variable. If a unique relationship between the variables exists or can be postulated, the variables can be said to be *functionally related.* If measurements are used to approximate the form of a functional relationship, the only error will be due to random measurement error.

There is another type of relationship where, in addition to measurement error, there are errors due to real variation between individuals. An example is the relation of porosity to permeability. Different rock samples of the same porosity probably will have widely different permeabilities, though on the average there will be some kind of relation between permeability and porosity. Such relations are not unique but are confined to a single population. This type of relationship can be described by a *regression line* which delineates the central tendency of the data. Regression techniques can be used to define either functional or regression lines, but the latter type is probably the more prevalent in petroleum engineering problems.

In the regression analysis one variable is singled out as being dependent upon the other. There are some problems involving interdependent variables where neither variable can be specified as *dependent* or *independent.* In this situation the problem would not be to define the relationship (as this would be meaningless) but to determine whether a significant relationship exists. This would require a *correlation analysis.*

Correlation analysis is used where the main interest is in the distribution of the variables and in the association between them; whereas, the regression analysis is concerned with estimating one variable from another.

Correlation coefficient

In looking for correlation between two variables x_1 and x_2, the investigator takes a sample from a population and measures x_1 and x_2 on each individual in the sample.

The *correlation coefficient,* r, is given by

$$r = \Sigma(x_1 - \overline{X}_1)(x_2 - \overline{X}_2)/[\Sigma(x_1 - \overline{X}_1)^2\Sigma(x_2 - \overline{X}_2)^2]^{1/2} \quad (4.1)$$

The square of r represents the fraction of the total variance removed by correlation. Equation 4.1 may have values ranging from −1.0 to 1.0. A value of 0 is equivalent to no correlation and 1.0 or −1.0 represents a perfect correlation. The significance of a correlation can be tested by comparison of the computed r^2 with the standardized values in Table A-6. Table A-6 is utilized in much the same way as the tables of t, F and chi-square distribution. The investigator computes a sampling value of r^2 from the data using Equation 4.1 or 4.2. He enters Table A-6 at the degrees of freedom associated with r^2 and compares computed r^2 to the tabled values. If the computed value exceeds the tabled value at the 0.95 confidence level but not at the 0.99 confidence level, the correlation is considered significant. A significant correlation coefficient is flagged with an asterisk superscript. If the computed value exceeds the tabled value at the 0.99 confidence level, the correlation is considered highly significant and the correlation coefficient is flagged with a double asterisk superscript. Of course, if the computed correlation coefficient is less than the tabled value at the 0.95 confidence level, the correlation is considered nonsignificant. An alternate for Equation 4.1 which will make the computations easier is

$$r = [\Sigma x_1x_2 - (\Sigma x_1\Sigma x_2)/N]/\{[\Sigma x_1^2 - (\Sigma x_1)^2/N][\Sigma x_2^2 - (\Sigma x_2)^2/N]\}^{1/2} \quad (4.2)$$

The following problem illustrates application of the correlation coefficient. The question is "do the data of Table 4A suggest a correlation between size of investment and rate of return?"

TABLE 4A

x_1 = Capital investment, MM $	x_2 = Rate of return, %	x_1^2	x_2^2	x_1x_2
1.000	22.25	1.0000	495.062	22.250
0.820	23.12	0.6724	534.534	18.958
0.740	19.08	0.5476	364.046	14.119
0.690	20.17	0.4761	406.829	13.917
0.550	18.06	0.3025	326.164	9.933
0.475	12.13	0.2256	147.137	5.762
0.425	14.95	0.1806	223.502	6.354
0.380	10.60	0.1444	112.360	4.028
0.340	15.20	0.1156	231.040	5.168
0.310	16.00	0.0961	256.000	4.960
5.730	171.56	3.7609	3096.674	105.449

Applying equation 4.2,

$$
\begin{aligned}
r &= [105.449 - (5.730)(171.56)/10]/\{[3.7609 - (5.730)^2/10] \\
&\quad [3096.674 - (171.56)^2/10]\}^{1/2} \\
&= 0.834 \\
r^2 &= 0.696^{**}
\end{aligned}
$$

Table A-6 gives the values of r^2 for linear correlation of two variables at the 95% and 99% significance levels. If there is correlation, r^2, the fraction of variance removed by the correlation, will be as large or larger than indicated by Table A-6, indicating that there would be only a 5% chance at the 95% level or a 1% chance at the 99% level of getting this good a correlation when there was in fact no correlation. Table A-6 should be entered at N-2 degrees of freedom. In correlation analysis N represents the number of pairs of observations.

The correlation coefficient of the example problem is seen to be significant at the 99% level, and it has been flagged with two asterisks to denote this.

Rapid correlation methods

Correlation (or lack of it) often can be inferred from a *scatter diagram.* This diagram does not require any computations and is especially useful for large samples. The scatter diagram is simply a plot of one variable against the other. It is customary in plotting data to have the ordinate represent the dependent variable and the abcissa the independent variable, but this connotation does not hold for the scatter diagram because neither variable is classifiable as "dependent" or "independent."

As an example of the use of a scatter diagram, consider the following problem. The question has been asked, "Within a certain geological basin, is there any correlation between depth and API gravity of the oil produced?" The engineer consults scout reports and finds records of 43 discovery wells which report depth of producing interval and the API gravity of the produced oil. These data are plotted as a scatter diagram, which is shown in Fig. 4.1. An ellipse encompassing the data shows, qualitatively, the amount of correlation. Obviously, a highly

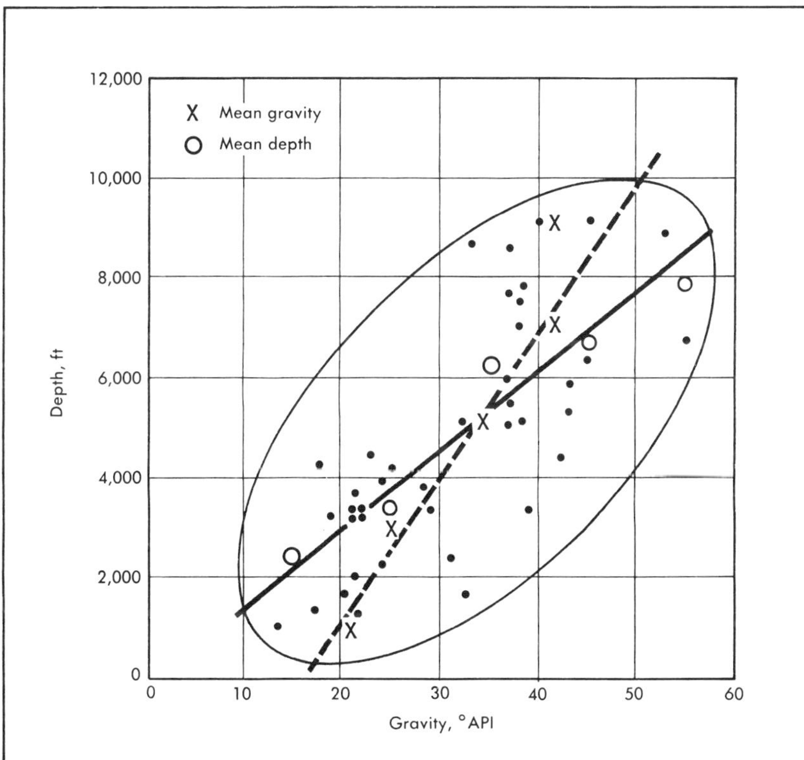

FIG. 4.1 Scatter diagram showing correlation of API gravity and depth of some Gulf Coast discovery wells.

elongate ellipse would indicate a high degree of correlation while a less elongate ellipse would indicate low correlation. Correlations are *positive* when the slope of the major axis of the ellipse is positive and *negative* when the slope is negative. The example problem displays a positive correlation; that is, as depth increases, the API gravity of the oil also increases.

If the investigator wants to go a step farther and get a semi-quantitative assessment of the significance of the correlation, there is a simple graphical technique he can apply. It is based on a property of regressions which will be elaborated upon later. This property is that the correlation coefficient is given by the square root of the ratio of slopes of the two regression lines: the regression of x_1 upon x_2 and the regression of x_2 upon x_1. The procedure is:

1) Divide the x_1 and x_2 axes of the scatter diagram into a convenient number of class intervals.

2) Replace the points in each x_1 interval by a class mean and fit the best straight line through the means by eye.

3) Replace the points in each x_2 interval by a class mean and fit the best straight line through the means by eye.

4) Compute the slope, $b = dx_2/dx_1$, of each line and calculate $r^2 = b/b'$.

5) Test r^2 for significance.

This technique has been applied to the data of Fig. 4.1. The scatter diagram was partitioned into depth classes of 2,000 ft and gravity classes of 10° API. The mean of the points within each interval was calculated and plotted at the midpoint of the interval and the two regression lines were fitted by eye. Slopes of these lines are 160 and 294, so $r^2 = 160/294 = 0.544$. There are 43 data points; therefore, the degrees of freedom associated with r are $N - 2 = 41$. Table A-6 shows that the correlation is highly significant.

The *rank correlation test* is useful when the number of observations is small. In this test a variable x_1 is ranked in some order and then the variable x_2 is also ranked for each of the items in the first series. A simple procedure is then followed to determine whether there is a significant correlation between the two sets of rankings. A tabulation of the distribution of rank correlation coefficients is shown in Table A-7. The investigator computes the rank correlation coefficient from the sample using Equation 4.3. The computed R is compared to the values of Table A-7 in the same way that Table A-6 is used for r^2.

Table A-7 is slightly different in that one enters it at N (the number of pairs of observations) instead of df (the degrees of freedom). If R has a positive value, the correlation is positive. Likewise, a negative value of R denotes that the correlation is negative.

Table 4B is the rate of return data of the previous problem arranged for a rank correlation test.

TABLE 4B

Capital investment, M$	Rate of return, %	Investment rank	Rate of return rank	Rank difference, d
1000	22.25	1	2	1
820	23.12	2	1	1
740	19.08	3	4	1
690	20.17	4	3	1
550	18.06	5	5	0
475	12.13	6	9	3
425	14.95	7	8	1
380	10.60	8	10	2
340	15.20	9	7	2
310	16.00	10	6	4

From the rank differences the rank correlation coefficient is calculated by

$$R = 1 - 6\Sigma d^2/N(N^2 - 1) \tag{4.3}$$

For the Table 4B data

$$R = 1 - (6)(38)/(10)(99) = 0.770^{**}$$

If R is tested in Table A-7, it will be found significant at the 99% probability level. Thus it can be said with less than 1% probability of error that there is a positive correlation between investment and rate of return.

If two items should have the same rank, assign the average of the two to both items. For example, if two items are tied for fifth place the ranking would be 3, 4, 5.5, 5.5, 7, 8, etc.

Linear regression

Physical and chemical laws can usually be expressed in mathematical form. Some of these have been derived empirically and others have been derived mathematically. *Regression analysis* deals with the problem of deriving an *empirical* relation between variables.

If it is known that y varies linearly with x, then an approximation to the form of this relationship over given ranges of the variables can be estimated from measurements of the corresponding values of y and x when they are varied over the given ranges.

Because of the usual variation associated with data, the plotted points of a scatter diagram of y versus x will not fall exactly on a line; but they will vary randomly around one. The line representing the average of an infinite number of determinations would be the graph of the relation. It is the task of regression analysis to provide an approximation of that line from a limited number of determinations and to provide confidence limits for the approximation.

It is conventional for the dependent variable to be designated y and plotted along the ordinate and the independent variable to be designated x and plotted along the abcissa. The regression line is the line which minimizes variation in the y direction, only. It therefore assumes that there is no variation in the x direction. The line is known as a *regression of y upon x.* Since there are always errors in measured values of x as well as y, an entirely different relation may result if the variables are reversed and the regression of x upon y is calculated. For this reason a regression of y upon x should not be used to predict a specific value of x, and a regression of x upon y should not be used to predict a specific value of y. As might be expected, the two regressions provide a basis for determining the correlation coefficient. This will be illustrated later.

There are several methods that could be applied to get a central tendency line through scattered data points, such as:

1) Deviation in the y direction can be minimized.
2) Deviation in the x direction can be minimized.
3) Deviation perpendicular to the line can be minimized.
4) The squares of (1), (2) or (3) can be minimized.

The method of least squares on y is the most commonly used and is the subject of the regression methods described herein.

Consider a relation y = F(x). If it is a straight-line relationship, it may be written

$$\hat{y} = b_o + b_1 x \tag{4.4}$$

where $\hat{y}$ is an estimate of y at any fixed value of x. Equation 4.4 represents the regression of y upon x.

The least squares procedure specifies that the regression line must satisfy the condition: that the sum of the squares of all deviations of estimated y from actual y be minimized. Therefore, $\sum_{i=1}^{N} (y_i - \hat{y}_i)^2 \rightarrow$ minimum. According to Equation 4.4 $\hat{y}_i = b_o + b_1 x_i$, so it may be stated that $\sum_{i=1}^{N} (y_i - b_o - b_1 x_i)^2 \rightarrow$ minimum. Expanding the expression for the sum of squares of the differences produces the expression

$$\sum_{i=1}^{N} [y_i^2 - 2b_o y - 2b_1 xy + b_o^2 + 2b_o b_1 x + b_1^2 x^2]$$

Dropping the subscripts for the sake of convenience, the summation is

$$\Sigma y^2 - 2b_o \Sigma y - 2b_1 \Sigma xy + Nb_o^2 + 2b_o b_1 \Sigma x + b_1^2 \Sigma x^2 \qquad (4.5)$$

When the sum of squares is a minimum, the derivatives with respect to b_o and b_1 will be zero, so

$$\partial\Sigma(y - \hat{y})^2/\partial b_o = 0 = -2\Sigma y + 2Nb_o + 2b_1 \Sigma x \qquad (4.6)$$

and

$$\partial\Sigma(y - \hat{y})^2/\partial b_1 = 0 = -2\Sigma xy + 2b_o \Sigma x + 2b_1 \Sigma x^2 \qquad (4.7)$$

According to Equation 4.6

$$b_o = \Sigma y/N - b_1 \Sigma x/N = \overline{Y} - b_1 \overline{X} \qquad (4.8)$$

also, according to Equation 4.7

$$b_o = \Sigma xy/\Sigma x - b_1 \Sigma x^2/\Sigma x \qquad (4.9)$$

After the sum of squares of the deviations have been minimized, there will remain a residual error, e; therefore, the regression equation should, in reality, be written $\hat{y} = b_o + b_1 x + e$. One may find the expression for the residual sum of squares by gathering and rearranging the terms of Equation 4.5 as follows:

$$\Sigma(y - \hat{y})^2 = [\Sigma y^2 - b_o \Sigma y - b_1 \Sigma xy] - [b_o(\Sigma y - b_o - b_1 \Sigma x)]$$
$$-[b_1 \Sigma x(y - b_o - b_1 \Sigma x)]$$

The second and third terms are zero, leaving

$$\text{Residual sum of squares} = \Sigma y^2 - b_o\Sigma y - b_1\Sigma xy$$

Substituting Equation 4.8 leads to

$$\text{Residual s.s.} = \Sigma y^2 - (\Sigma y)^2/N - b_1(\Sigma xy - \Sigma x\Sigma y/N) \quad (4.10)$$

When Equations 4.8 and 4.9 are solved simultaneously, the following equations for b_o and b_1 result.

$$b_o = [\Sigma x^2\Sigma y - (\Sigma xy)(\Sigma x)]/N\Sigma x^2 - (\Sigma x)^2] \quad (4.11)$$

$$b_1 = (N\Sigma xy - \Sigma x\Sigma y)/[N\Sigma x^2 - (\Sigma x)^2] \quad (4.12)$$

The constant, b_o, is the y-intercept of the regression line and b_1 is the slope. The constant, b_1, is called the *regression coefficient* of y upon x.

A characteristic of the regression is that it always passes through $(\overline{X}, \overline{Y})$, the arithmetic mean of all the data. Thus, $(\overline{X}, \overline{Y})$ and b_1 define the regression line. This is illustrated by Equation 4.8. Three individual sources of variance contribute to the straight-line regression. These are: variance about the mean, variance due to the slope and residual variance.

The residual variation, represented by the deviations of the y-observations from the line, must be less than the total variation of the y-observations about their mean, and the extent by which it is less represents the amount of variation accounted for by the slope. It will be recalled from "analysis of variance" that the sums of squares of variance due to various sources are additive; therefore, the sums of squares contributing to the linear regression can be summarized as

$$\Sigma(y - \hat{y})^2 = \Sigma(y - \overline{Y})^2 - \Sigma(\hat{y} - \overline{Y})^2$$

where

$\Sigma(y - \hat{y})^2$ = residual sum of squares
$\Sigma(y - \overline{Y})^2$ = sum of squares about the mean
$\Sigma(\hat{y} - \overline{Y})^2$ = sum of squares due to the slope

The residual sum of squares is represented by Equation 4.10. The sum of squares about the mean is $\Sigma y^2 - (\Sigma y)^2/N$, the first two terms of Equation 4.10. Also, it will be recognized as Equation 3.5, the "sum of squares within samples" from analysis of variance. The last term of Equation 4.10, $b_1(\Sigma xy - \Sigma x\Sigma y/N)$, is the sum of squares due to the slope. It may also be written as $b_1^2\ [\Sigma x^2 - (\Sigma x)^2/N]$.

The residual sum of squares divided by degrees of freedom, N − 2, is an estimate of σ_e^2, the *variance about the regression,* often referred to as *error variance.* The *standard error of estimate* is

$$s_e = [(\text{residual s.s.})/(N - 2)]^{1/2} \tag{4.13}$$

The variance of the regression coefficient b_1 is given by

$$s^2(b_1) = s_e^2/\Sigma(x - \overline{X})^2 = s_e^2/[\Sigma x^2 - (\Sigma x)^2/N] \tag{4.14}$$

and the confidence limits on the estimate of b_1 are

$$b_1 \pm t_{1/2\alpha}s(b_1) \tag{4.15}$$

where

$t_{1/2\alpha}$ = the value of t at $1/2\alpha$ probability level (Table A-1). and N − 2 degrees of freedom.

In any linear regression problem, if the confidence limits on the slope do not include zero, the slope can be considered statistically significant at the probability level in question.

The variance of the regression estimate $\hat{y}_i$ is given by

$$s^2(\hat{y}_i) = s^2(\overline{Y}) + (x_i - \overline{X})^2\, s^2\,(b_1)$$

This expresses the variance of y about $\hat{y}$ at any given value of x. The individual elements will be recognized as the variance of the mean and the variance of the slope. The variance of the mean is s_e^2/N, and the variance of the slope is given by Equation 4.14 so when these are combined,

$$\begin{aligned} s_i^2(\hat{y}_i) &= s_e^2/N + (x_i - \overline{X})^2\, s_e^2/\Sigma(x_i - \overline{X})^2 \\ &= s_e^2[1/N + (x_i - \overline{X})^2/\Sigma(x - \overline{X})^2] \end{aligned} \tag{4.16}$$

It can be seen from Equation 4.16 that the error in $\hat{y}$ is a minimum at $x_i = \overline{X}$ and increases as x_i deviates farther from $\overline{X}$. This fact points up the hazards of indiscriminate extrapolation of regressions, because as one moves farther from the centroid of the data, $(\overline{X}, \overline{Y})$, the probable error in $\hat{y}$ gets greater. This is illustrated graphically by Fig. 4.2.

The confidence limits on $\hat{y}_i$ are

$$\hat{y}_i \pm t_{1/2\alpha}s(\hat{y}_i) \tag{4.17}$$

Equation 4.16 applies to any point included in the data from which the regression was obtained. If the regression is to be used to predict

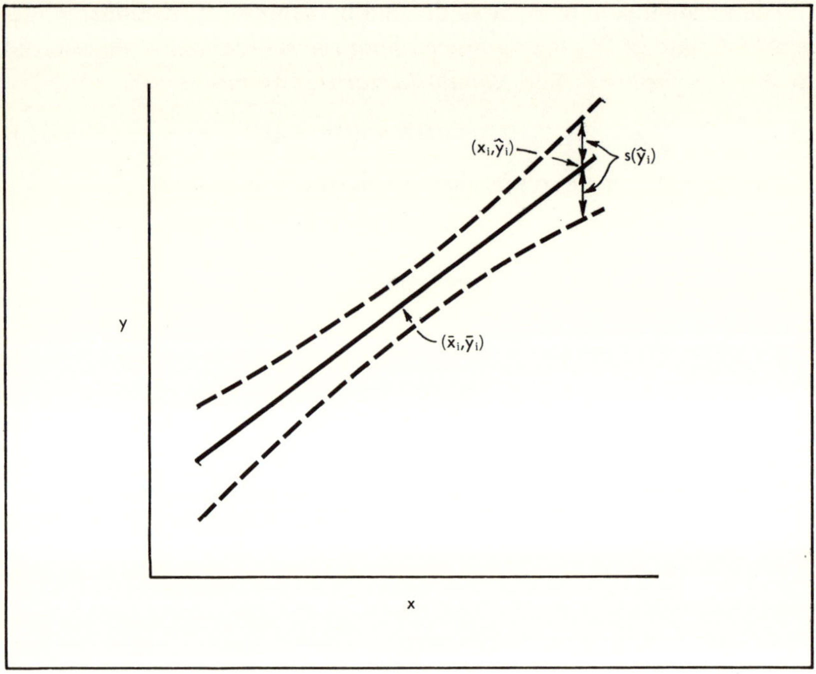

FIG. 4.2 Diagram showing a linear regression and showing the trend of the standard error of an estimate of y.

what $\hat{y}_i$ would be if a new set of data points were obtained, then the variance about the regression, s_e^2, would have to be added, and Equation 4.16 would become

$$s^2(\hat{y}_i') = s_e^2[1 + 1/N + (x_i - \overline{X})^2/\Sigma(x - \overline{X})^2]$$

One other statistic that can be obtained from the regression data is the correlation coefficient r. Under "Rapid Correlation Methods" the approximate square of the correlation coefficient was expressed as the ratio of slopes of the two regression lines "y upon x" and "x upon y." The correlation coefficient is, in fact, derived on this basis. If we define the correlation coefficient as the square root of the ratio of the slopes, it can be written as:

$$r = \sqrt{bb'}$$

where b is the slope, dy/dx, of the regression of y upon x; and b′ is the slope, dx/dy, of the regression of x upon y. The slope of the regression of y upon x is the regression coefficient b_1 which has just been derived.

Equation 4.12 is an alternate form of:

$$b_1 = \Sigma(x - \overline{X})(y - \overline{Y})\ \Sigma(x - \overline{X})^2.$$

(The alternate forms for sums of squares have been presented previously in Equations 3.4 through 3.10 under "Analysis of Variance.")

The slope of the regression of x upon y can be expressed as:

$$b' = \Sigma(y - \overline{Y})(x - \overline{X})/\Sigma(y - \overline{Y})^2$$

simply by interchanging the x's and y's in the slope equation. Substituting for b and b′ the equation for correlation coefficient is:

$$r = \Sigma(x - \overline{X})(y - \overline{Y})/[\Sigma(x - \overline{X})^2\Sigma(y - \overline{Y})^2]^{1/2}$$

or alternately,

$$r = (\Sigma xy - \Sigma x\Sigma y/N)/\{[\Sigma x^2 - (\Sigma x)^2/N][\Sigma y^2 - (\Sigma y)^2/N]\}^{1/2}$$

which will be recognized as Equations 4.1 and 4.2, respectively, with x_2 being replaced by y.

The numerator of the correlation coefficient equation is a measure of the variability of the crossproduct. This crossproduct expresses the *covariance in x and y*. The first term of the denominator is the sum of squares about $\overline{X}$ and the second term of the denominator is the sum of squares about $\overline{Y}$. The correlation coefficient actually, then, is:

$$r = \text{cov}(xy)/s_x s_y$$

where:

$$\text{cov}(xy) = \Sigma(x - \overline{X})(y - \overline{Y})/(N - 1)$$
$$s_x = [\Sigma(x - \overline{X})^2/(N - 1)]^{1/2}$$
$$s_y = [\Sigma(y - \overline{Y})^2/(N - 1)]^{1/2}$$

The degrees of freedom, N − 1, are the same for each of the sums of squares, and the degrees of freedom appear both in numerator and denominator therefore they cancel, leading to the expanded expressions represented in Equations 4.1 and 4.2.

Squaring the abbreviated expression, we have

$$r^2 = [\text{cov}(xy)]^2/s_x^2 s_y^2$$

It has just been shown that

$$b_1 = \Sigma(x - \overline{X})(y - \overline{Y})/\Sigma(x - \overline{X})^2$$
$$= cov(xy)/s_x^2$$

Substituting, we have

$$r^2 = b_1 \; cov(xy)/s_y^2 = b_1\Sigma(x - \overline{X})(y - \overline{Y})/s_y^2$$

The expression can be simplified by multiplying numerator and denominator by s_x^2 so that

$$r^2 = b_1\Sigma(x - \overline{X})(y - \overline{Y})s_x^2/s_x^2 s_y^2$$
$$= b_1^2 s_x^2/s_y^2.$$

Expressed in sums of squares,

$$r^2 = b_1^2\Sigma(x - \overline{X})^2/\Sigma(y - \overline{Y})^2 \quad (4.18)$$

The square of the correlation coefficient represents the *fraction of the variance accounted for by the correlation between the variables.* In this context the mathematical equivalence of regression and correlation becomes obvious.

Matrix methods for regression analysis

The straight-line regression presents no particular computional problems, but regressions involving several independent variables or polynomials in a single independent variable can become quite complex. These problems can be solved systematically and relatively easily using matrix notation and a modest amount of matrix algebra. There is some advantage to applying the matrix approach to linear regression also, because it facilitates the partition of variances and it is the easiest way to become familiar with the matrix method.

All regression equations take the form

$$\hat{y} = b_0 + b_1x_1 + b_2x_2 + \cdot\cdot\cdot + b_px_p$$

In this equation x_1, x_2, x_3, etc. may be different variables or successive powers of the same variable. The first two terms will be recognized as the straight-line regression equation.

The least squares technique provides $p + 1$ equations which, when solved simultaneously, will yield the regression constants b_o, b_1, b_2, etc. The matrix method of solving linear simultaneous equations in p variables involves arranging the coefficients of the equations in tables or rectangular arrays which are called *matrices,* and applying matrix algebra to their *determinants.*

Appendix B contains a brief summary of the principles of matrix manipulation, and it describes the *Abbreviated Doolittle technique* for solving regression and correlation problems. The Abbreviated Doolittle method will be used in all the numerical solutions which follow.

Example problem – linear regression

Table 4C presents a set of production decline data which have been prepared for a constant percentage production decline curve. The constant percentage decline is $\log_{10}$ of production rate versus time; a characteristic of the constant percentage decline being that the $\log_{10}$ transformation of q_t plots as a straight line against time. The problem is to fit a least squares line to the data of Table 4C, and to place confidence limits upon the least-squares fit.

TABLE 4C

$x = t$, Time, months	$y = q_t$ prod. rate, Mbbl/month	$y' = \log_{10}q_t$	x^2	$(y')^2$	xy'
17	16.0	1.204	289	1.450	20.468
18	17.2	1.236	324	1.528	22.248
19	13.8	1.140	361	1.300	21.660
20	14.2	1.152	400	1.327	23.040
21	13.5	1.131	441	1.279	23.751
22	13.2	1.121	484	1.257	24.662
23	13.1	1.118	529	1.250	25.714
24	13.2	1.121	576	1.257	26.904
25	12.6	1.100	625	1.210	27.500
26	12.0	1.079	676	1.164	28.054
27	11.3	1.053	729	1.109	28.431
28	10.5	1.022	784	1.044	28.616
30	10.5	1.022	900	1.044	30.660
31	10.0	1.000	961	1.000	31.000
32	9.5	0.978	1,024	0.956	31.296
33	9.2	0.964	1,089	0.929	31.812
34	9.6	0.982	1,156	0.964	33.388
35	8.5	0.929	1,225	0.863	32.515
465		19.352	12,573	20.931	491.719

The elements of the matrix for the forward Doolittle solution will be

N	Σx	Σy
	Σx^2	Σxy

following the form set out in Table B-1 of Appendix B. Solution of the matrix is as follows:

				j =
(1)	18	465	19.352	502.352
(2)		12,573	491.719	13,529.719
(3)	18	465	19.352	502.352
(4)		25.833	1.075	27.908
(5)		560.655	−8.156	552.499
(6)		1	−0.0145	0.985

Line 1: $j_1 = 18 + 465 + 19.352 = 502.352$

Line 2: $j_2 = 465 + 12{,}573 + 491.719 = 13{,}529.719$

Line 4: $B_{12} = 18/18 = 1;\ B_{13} = 465/18 = 25.833$

$B_{1y} = 19.352/18 = 1.075$

$j_4 = 502.352/18 = 27.908$

Line 5: $A_{22} = 12{,}573 - (465)(25.833) = 560.655$

$A_{2y} = 491.719 - (465)(1.075) = -8.156$

$j_5 = 13{,}529.719 - (465)(27.908) = 552.499$

Line 6: $B_{23} = 560.655/560.655 = 1$

$B_{2y} = -8.156/560.655 = -0.0145$

$j_6 = 552.499/560.655 = 0.985$

Lines 4 and 6 contain the coefficients for b_o and b_1, respectively.

$$b_1 = -0.0145$$
$$b_o + (25.833)(-0.0145) = 1.075$$
$$b_o = 1.450$$

Therefore, the least squares regression equation for log production rate as a function of the total months the field has been producing is

$$\log_{10} \hat{q}_t = 1.450 - 0.0145t$$

One of the advantages of the forward Doolittle solution is the facility with which it permits separation and evaluation of the sources of variance. The sums of squares are

$$\text{total s.s.} = \Sigma y^2 = 20.931$$
$$\text{s.s. absorbed by } b_o = (\Sigma y)^2/N = A_{oy}B_{oy} = (19.352)(1.075) = 20.803$$
$$\text{s.s. absorbed by } b_1 = b_1(\Sigma xy - \Sigma x\Sigma y/N) = A_{1y}B_{1y} = (-8.156)(-0.0145) = 0.118$$

Sufficient information is now available to permit an analysis of variance to be made. Table 4D is the table of variance.

TABLE 4D

Source of variance	Sum of squares	df	Mean square	F
Σy^2	20.931	18		
b_o	20.803	1		
Mean	0.128	17		
b_1	0.118	1	0.118	$F_{1,16} = 190$ **
Residual	0.010	16	0.00062	

The analysis of variance is carried out in the same manner as described in Chapter 3. One degree of freedom is removed for each regression coefficient, and the F-test is used to test the significance of source of variance by comparing the mean square for that source to the residual mean square. In the production decline problem the slope, b_1, is highly significant as one would expect from looking at a scatter diagram of the data. The scatter diagram is shown in Fig. 4.3, along with the least squares regression line and the confidence limits on $\hat{y}$.

The variance of the regression coefficient b_1 is given by Equation 4.14 as

$$s^2(b_1) = 0.00062/[12{,}573 - (465)^2/18] = 1.11 \times 10^{-6}$$

and the 95% confidence limits on b_1 are

$$-0.0145 \pm 2.120\sqrt{1.11 \times 10^{-6}} = -0.0167 \text{ and } -0.0123$$

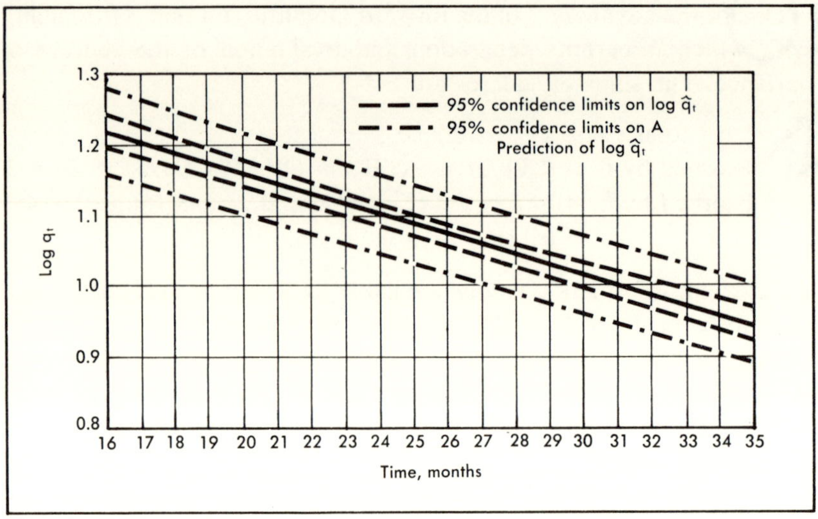

FIG. 4.3 Linear regression fitted to production-decline data.

Thus, there is 95% probability that the true value of the slope lies between −0.0123 and −0.0167; and the slope is significant because the limits do not include zero.

The variance of regression estimates $\hat{y}_i$ is given by Equation 4.16. For example at x = 34,

$$s^2(\hat{y}) = 0.00062\{1/18 + (34 - 25.833)^2/[12{,}573 - (465)^2/18]\}$$
$$= 0.000108$$

at x = 34, $\hat{y} = 1.450 - (0.0145)(34) = 0.957$, and the 95% confidence limits on $\hat{y}$ are

$$0.957 \pm 2.120\ \sqrt{1.08 \times 10^{-4}} = 0.935 \text{ and } 0.979$$

The 95% confidence limits on $\hat{y}$ and predictions of $\hat{y}$ are plotted in Fig. 4.3. The divergence of confidence limit lines from $(\overline{X}, \overline{Y})$ to the outer limits of the data is very slight, which is fortunate, because the purpose of a decline curve analysis is to extrapolate the curve. There is some theoretical justification for this extrapolation because the primary production rate always has some sort of functional relationship to time.

The square of the correlation coefficient is given by Equation 4.18 as

$$r^2 = (-0.0145)^2[12{,}573 - (465)^2/18]/[20.931 - (19.352)^2/18]$$
$$= 0.940^{**}$$

which, of course, is highly significant.

Example problem—curvilinear regression

In most data-fitting problems, it is preferable to seek a transformation of variables which will produce a straight-line regression, but it is not always possible to find a suitable transformation. When this hap-

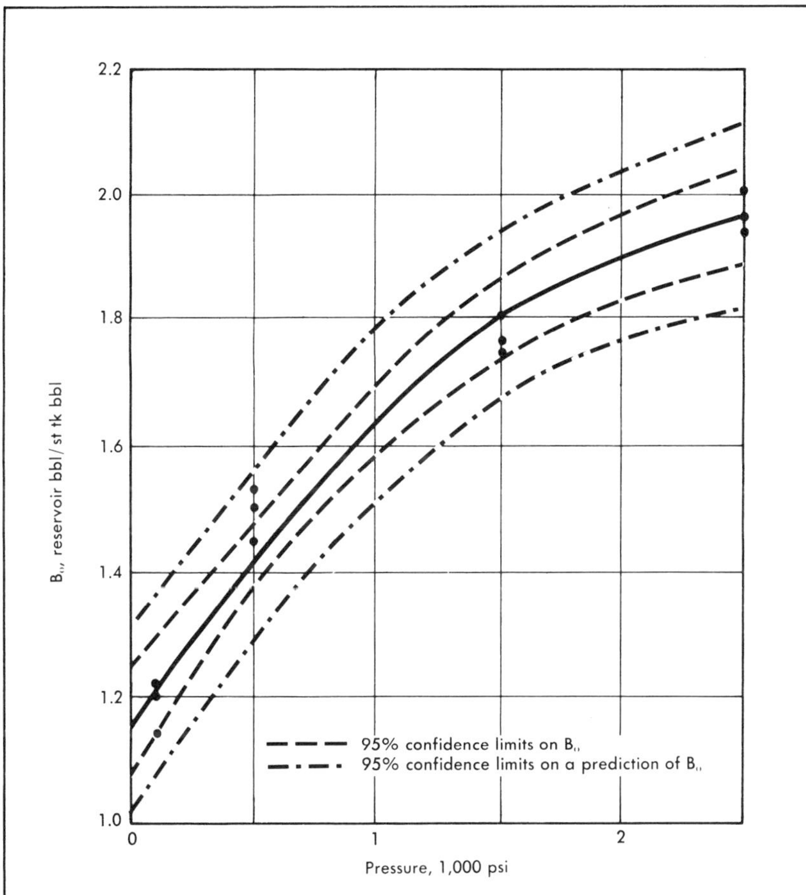

FIG. 4.4 Polynomial regression fitted to relative-volume-factor data.

pens it may be desirable to fit a polynomial to the data. The polynomial, which is simply a regression involving exponentials of x, takes the form

$$\hat{y} = b_o + b_1x + b_2x^2 + \cdots + b_px^p$$

The order of the polynomial to be used will depend upon the data. Significance tests on the partitioned variances will show whether a given polynomial provides an adequate fit of the data.

The method of computing curvilinear regressions by the forward Doolittle matrix is an extension of the linear regression technique and has been outlined in Appendix B. The elements of the matrix for polynomial regression differ from those of multivariate regression only in that the variable x_2 becomes x^2, x_3 becomes x^3, etc.

As an example of a curvilinear regression problem, consider Fig. 4.4 which shows a set of measurements of formation volume factor at various reservoir pressures. The problem is to fit the best curve to the data. A second-order fit will be tried, so the form of the equation will be

$$\hat{B}_o = b_o + b_1P + b_2P^2$$

The elements of the matrix for the forward Doolittle solution will be

N	Σx	Σx^2	Σy
	Σx^2	Σx^3	Σxy
		Σx^4	Σx^2y

TABLE 4E

$x = P$ (M psi)	$y = B_o$ (Res bbl/ S.T. bbl)	$x^2 = P^2$	$x^3 = P^3$	$x^4 = P^4$	$xy = PB_o$	$x^2y = P^2B_o$
0.100	1.22	0.0100	0.00100	0.000100	0.122	0.0122
0.100	1.20	0.0100	0.00100	0.000100	0.120	0.0120
0.100	1.14	0.0100	0.00100	0.000100	0.114	0.0114
0.500	1.53	0.2500	0.12500	0.062500	0.765	0.3825
0.500	1.50	0.2500	0.12500	0.062500	0.750	0.3750
0.500	1.44	0.2500	0.12500	0.062500	0.720	0.3600
1.500	1.80	2.2500	3.37500	5.062500	2.700	4.0500
1.500	1.76	2.2500	3.37500	5.062500	2.640	3.9600
1.500	1.74	2.2500	3.37500	5.062500	2.610	3.9150
2.500	2.00	6.2500	15.62500	39.062500	5.000	12.5000
2.500	1.96	6.2500	15.62500	39.062500	4.900	12.2500
2.500	1.94	6.2500	15.62500	39.062500	4.850	12.1250
13.800	19.23	26.2800	57.37800	132.562800	25.291	49.9531

Table 4E shows the computation of the matrix components. It will be noted that the products have not been rounded off, as it is essential to maintain a high level of precision during the matrix manipulations.

The matrix solution is:

					j =
(1)	12	13.8000	26.2800	19.2300	71.3100
(2)		26.2800	57.3780	25.2910	122.7490
(3)			132.5628	49.9531	266.1739
(4)	12	13.8000	26.2800	19.2300	71.3100
(5)	1	1.1500	2.1900	1.6025	5.9425
(6)		10.4100	27.1560	3.1765	40.7425
(7)		1	2.6086	0.3051	3.9138
(8)			4.1705	−0.4459	3.7218
(9)			1	−0.1069	0.8924

Line 1: $j_1 = 12 + 13.8000 + 26.2800 + 19.2300 = 71.3100$

Line 2: $j_2 = 13.8000 + 26.2800 + 57.3780 + 25.2910 = 122.7490$

Line 3: $j_3 = 26.2800 + 57.3780 + 132.5628 + 49.9531 = 266.1739$

Line 5: $12/12 = 1$; $13.8000/12 = 1.1500$; $26.2800/12 = 2.1900$
$19.2300/12 = 1.6025$; $j_5 = 71.3100/12 = 5.9425$

Line 6: $26.2800 - (13.8000)(1.1500) = 10.4100$
$57.3780 - (13.8000)(2.1900) = 27.1560$
$25.2910 - (13.8000)(1.6025) = 3.1765$
$j_6 = 122.7490 - (13.8000)(5.9425) = 40.7425$

Line 7: $10.4100/10.4100 = 1$; $27.1560/10.4100 = 2.6086$;
$3.1765/10.4100 = 0.3051$; $j_7 = 40.7425/10.4100 = 3.9138$

Line 8: $132.5628 - (26.2800)(2.1900) - (27.1560)(2.6086) = 4.1705$
$49.9531 - (26.2800)(1.6025) - (27.1560)(0.3051) = -0.4459$
$j_8 = 266.1739 - (26.2800)(5.9425) - (27.1560)(3.9138)$
$= 3.7218$

Line 9: $4.1705/4.1705 = 1$; $-0.4459/4.1705 = -0.1069$
$j_9 = 3.7218/4.1705 = 0.8924$

The regression coefficients are

$$b_2 = -0.1069$$
$$b_1 = 0.3051 - (2.6086)(-0.1069) = 0.5840$$
$$b_0 = 1.6025 - (2.1900)(-0.1069) - (1.1500)(0.5840)$$
$$= 1.1650$$

so the regression equation is

$$\hat{B}_o = 1.1650 + 0.5840P - 0.1069P^2$$

The sums of squares are

Total s.s. $= \Sigma y^2 = 31.8629$
s.s. absorbed by $b_o = A_{oy}\, B_{oy} = (19.2300)(1.6025) = 30.8161$
s.s. absorbed by $b_1 = A_{1y}\, B_{1y} = (3.1765)(0.3051) = 0.9692$
s.s. absorbed by $b_2 = A_{2y}\, B_{2y} = (-0.4459)(-0.1069) = 0.0477$

In the partitioning of variances in any regression problem, the residual variance can be partitioned farther into "lack of fit" and "experimental error" if an estimate of the experimental error is available. In a problem such as this involving laboratory data, experimental error often is known or can be determined. The estimate can be made from the regression data here because several independent measurements of y were made at each constant value of x. At any value of x the sum of squares of experimental error is $\Sigma y^2 - (\Sigma y)^2/N$, and the degrees of freedom are $N - 1$. The total of the individual sums of squares is the total error sum of squares, for example.

At x = 0.100; s.s. = 4.2280 − 4.2245 = 0.0035; df = 2
At x = 0.500; s.s. = 6.6645 − 6.6603 = 0.0042; df = 2
At x = 1.500; s.s. = 9.3652 − 9.3633 = 0.0019; df = 2
At x = 2.500; s.s. = 11.6052 − 11.6033 = 0.0019; df = 2
total s.s. = 0.0115; total df = 8

If it had been necessary, the experimental error could have been obtained from an independent experiment or an estimate, *provided the experimental error, itself, is not a function of x.*

When the experimental error is derived from an independent source, the degrees of freedom associated with it is one. The sum of squares due to lack of fit is the residual sum of squares minus the sum of squares due to experimental error. Likewise, the degrees of freedom associated with the estimate of lack of fit is equal to the degrees of freedom associated with the residual minus the degrees of freedom associated with the estimate of experimental error.

Table 4F is the analysis of variance table.

The F-test shows that the coefficients b_1 and b_2 are highly significant and the lack of fit is significant, indicating that a better fit could be gained by going to the next higher polynomial.

TABLE 4F

Source of variance	Sum of squares	df	Mean square	F
Σy^2	31.8629	12		
b_o	30.8161	1		
Mean	1.0468	11		
b_1	0.9692	1	0.9692	$F_{1,8} = 692$ **
b_2	0.0477	1	0.0477	$F_{1,8} = 34.1$ **
Residual	0.0299	9	0.0033	
lack of fit	0.0184	1	0.0184	$F_{1,8} = 13.1$ *
Experimental Error	0.0115	8	0.0014	

This completes the information that can be gained from the forward Doolittle solution. If the investigator requires only the regression equation and the analysis of variance, the forward solution is sufficient; but if he needs the confidence limits on $\hat{y}$ and on the regression coefficients, it will be necessary to calculate the inverse matrix, C.

The variances of the regression coefficients are given by the product of the residual variance and the appropriate diagonal element of the inverse matrix, thus

$$s^2(b_1) = s_e^2 c_{11}$$
$$s^2(b_2) = s_e^2 c_{22}$$
$$s^2(b_3) = s_e^2 c_{33}, \text{ etc.}$$

The nondiagonal terms of the inverse matrix give, similarly, the covariances of the estimates, such as

$$\text{cov}\,(b_1 b_2) = s_e^2 c_{12} = s_e^2 c_{21}$$

The covariances of the estimates indicate the extent to which chance errors in the estimate of any one coefficient will affect the estimates of other coefficients; they reflect any correlations between observed values of the independent variables. The covariance has little meaning in a curvilinear regression problem because each "independent" variable is simply another exponential of x.

Confidence limits on the regression coefficients are calculated from the variances by use of the appropriate t-multipliers (see Chapter 2, "Confidence Limits") thus the confidence limits on regression coefficient b_i are

$$b_i \pm t_{1/2\,a} s_e \sqrt{c_{ii}} \tag{4.19}$$

where t has the same number of degrees of freedom as s_e^2.

The precision with which b_i is estimated depends upon the magnitude of s_e^2, that is, on the residual scatter about the regression, on the number of observations available and on the inverse matrix element c_{ii}. This last depends on the spread of the observed x_i values and also on the extent to which the variations in x_i are correlated with variations in the other "independent" variables.

The matrix is inverted by the so-called "backward" Doolittle solution. The general procedure has been outlined in Appendix B on "Matrix Methods." The matrix of A's and B's calculated from the forward Doolittle procedure are shown schematically below.

A_{00}	A_{01}	A_{02}
	B_{01}	B_{02}
	A_{11}	A_{12}
		B_{12}
		A_{22}

The procedure is to calculate the elements of the C matrix by inversion of the A matrix, as follows:

$$c_{22} = 1/A_{22} = 1/4.1705 = 0.2398$$
$$c_{12} = -B_{12}c_{22} = (-2.6086)(0.2398) = -0.6255$$
$$c_{02} = -B_{01}c_{12} - B_{02}c_{22} = (-1.1500)(-0.6255) - (2.1900)(0.2398) = 0.1941$$
$$c_{11} = 1/A_{11} - B_{12}c_{12} = 1/10.4100 - (2.6086)(-0.6255) = 1.7278$$
$$c_{01} = -B_{01}c_{11} - B_{02}c_{12} = (-1.1500)(1.7278) - (2.1900)(-0.6255) = -0.6172$$
$$c_{00} = 1/A_{00} - B_{01}c_{01} - B_{02}c_{02} = 1/12 - (1.1500)(-0.6172) - (2.1900)(0.1941) = 0.3680$$

A quick check on the calculations is provided by the products, $\Sigma a_{ij}c_{ij}$, where a_{ij} is an element of the normal equation matrix and c_{ij} is the corresponding element of the inverted matrix. The column sums (equivalent to the diagonal elements of the identity matrix, I) of $a_{ij}\,c_{ij}$ should be equal to one. This refers to the complete matrices, not just the forward portions.

In the example problem,

$$a_{02}c_{02} + a_{12}c_{12} + a_{22}c_{22} = (26.2800)(0.1941) + (57.3780)(-0.6255) + (132.5628)(0.2398) = 0.9996$$

$$a_{01}c_{01} + a_{11}c_{11} + a_{21}c_{21} = (13.8000)(-0.6172) + (26.2800)(1.7278) + (57.3780)(-0.6255) = 0.9993$$

$$a_{00}c_{00} + a_{10}c_{10} + a_{20}c_{20} = (12)(0.3680) + (13.8000)(-0.6172) + (26.2800)(0.1941) = 0.9996$$

Information is now sufficient for calculating the confidence limits on the estimates of b_1 and b_2, using Equation 4.19.

The 95% confidence limits on b_1 at 9 degrees of freedom are

$$0.5840 \pm 2.262\ \sqrt{0.0033 \times 1.7278} = 0.4132 \text{ and } 0.7548$$

and the 95% confidence limits on b_2 at 9 degrees of freedom are

$$-0.1609 \pm 2.262\ \sqrt{0.0033 \times 0.2398} = -0.0973 \text{ and } -0.2245$$

It can be seen from the confidence limits that neither coefficient is very precisely determined. The F-test showed that b_1 and b_2 are highly significant and also that the residual variance, s_e^2, contains a significant amount of lack of fit. It is obvious that addition of a third term, b_3x^3, would reduce s_e^2 and consequently improve the estimates of all of the regression coefficients.

Finally, the variance of a regression estimate, $\hat{y}_i$, is given by

$$\begin{aligned} s^2(\hat{y}) &= s_e^2[1/N + c_{ii}(x_i - \overline{X}_i)^2 + c_{jj}(x_j - \overline{X}_j)^2 \\ &\quad + 2c_{ij}(x_i - \overline{X}_i)(x_j - \overline{X}_j)] \\ &= s_e^2/N + s^2(b_i)(x_i - \overline{X}_i)^2 + s^2(b_j)(x_j - \overline{X}_j)^2 \\ &\quad + 2\ \mathrm{cov}\ (b_ib_j)(x_i - \overline{X}_i)(x_j - \overline{X}_j) \end{aligned} \tag{4.20}$$

The confidence limits on $\hat{y}_i$ are given by Equation 4.17, where the degrees of freedom are the residual degrees of freedom, $N - p - 1$.

As in the case of the straight-line regression, s_e^2 must be added to $s^2(\hat{y}_i)$ if the confidence limits on a predicted value of $\hat{y}$ are required.

For example, at $P = 1$,

$$\hat{B}_o = 1.1650 + (0.5840)(1) - (0.1069)(1)^2 = 1.6421$$

The variances of the estimates of regression coefficients are

$$\begin{aligned} s^2(b_1) &= s_e^2c_{11} = (0.0033)(1.7278) = 0.00570 \\ s^2(b_2) &= s_e^2c_{22} = (0.0033)(0.2398) = 0.00079 \\ \mathrm{cov}(b_1b_2) &= s_e^2c_{12} = (0.0033)(-0.6255) = -0.00206 \end{aligned}$$

and the variance of $\hat{y}$ is

$$\begin{aligned} s^2(\hat{B}_o) &= s_e^2/N + s^2(b_1)(P - \bar{P})^2 + s^2(b_2)(P^2 - \bar{P}^2)^2 \\ &\quad + 2\,\mathrm{cov}(b_1b_2)(P - \bar{P})(P^2 - \bar{P}^2) \\ &= (0.0033/12) + (0.00570)(1 - 1.15)^2 + (0.00079)(1 - 2.19)^2 \\ &\quad + (2)(-0.00206)(1 - 1.15)(1 - 2.19) \\ &= 0.000787 \end{aligned}$$

The 95% confidence limits on $\hat{B}_o$ are

$$1.6421 \pm 2.262\ \sqrt{7.87 \times 10^{-4}} = 1.5786 \text{ and } 1.7056$$

The variance of a predicted $\hat{B}_o$ is

$$s^2(\hat{B}_o) + s_e^2 = 0.000787 + 0.0033 = 0.00409$$

and the 95% confidence limits are

$$1.6431 \pm 2.262\ \sqrt{40.9 \times 10^{-4}} = 1.4986 \text{ and } 1.7876$$

Fig. 4.4 shows the data points, the regression curve and the confidence limits on $\hat{B}_0$ and predicted $\hat{B}_0$.

Regressions by graphical methods

An unbiased estimate of the regression of y upon x can be obtained by using results from preselected values of x. The mean of an array of observed y-values at each preselected value of x gives an unbiased estimate y for that x, and the locus of means defines the regression curve. If the engineer has an opportunity to plan the experiment so that several independent observations of y are made at each fixed x, he can often get a more precise definition of the regression curve than could be obtained from the same number of points scattered over the same range of the independent variable.

Though in most cases it is not possible for the petroleum engineer to specify the conditions under which his data will be obtained, he can get a useful approximation of the regression curve by grouping the x's and plotting $\bar{X}_i$ versus $\bar{Y}_i$. The confidence limits on $\bar{Y}_i$ are an approximation of the confidence limits on y_i, *provided the distribution of* y_i *is normal.*

The graphical technique is illustrated by Fig. 4.5 which is a regression of permeability upon porosity derived from a group of randomly-

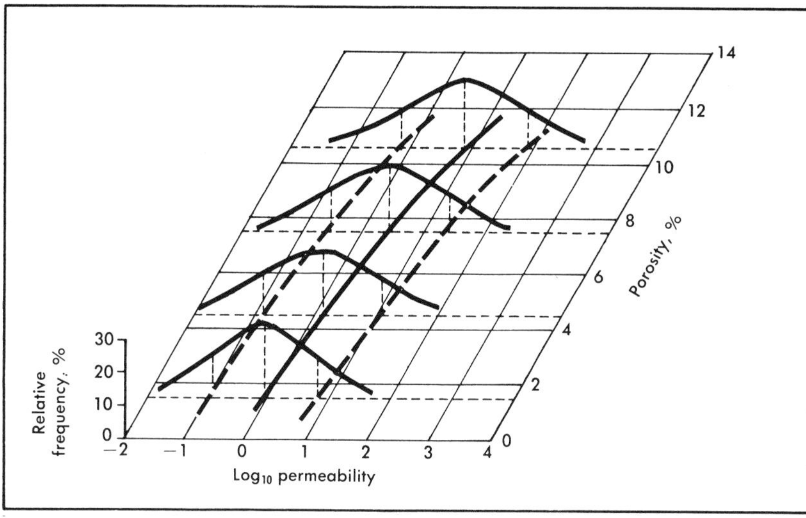

FIG. 4.5 A regression of permeability on porosity by graphical method.

selected core plugs from a tight carbonate reservoir. As is evident from the schematic presentation of Fig. 4.5, the samples were grouped into four porosity ranges whose averages are 1.5%, 4.5%, 7.5% and 10.5%. The distribution curves for $\log_{10}$ permeability were developed by differentiation of smoothed cumulative frequency curves plotted on normal probability paper (see Chapter I, "Graphical Methods"). Permeability was transformed to $\log_{10}$ K to get a more nearly normal distribution.

The regression curve is represented by the solid line connecting average log K within each porosity group. (In a normally distributed system this is also the median and the most frequent value of log K within each group). Standard error limits are shown by the shaded areas and are delineated by the dashed lines. These limits are the 16.7 and 83.3 percentile values of K within each porosity group. They should not be confused with standard error limits that would be obtained from a least squares calculation, because in the graphical method the standard error of each porosity group was developed independently; consequently the group variances are not necessarily related to the overall average, $(\overline{X}, \overline{Y})$, as in the analytical method. In other words, the graphical method would show any non-homogeneity of variances that might exist between porosity groups, while the

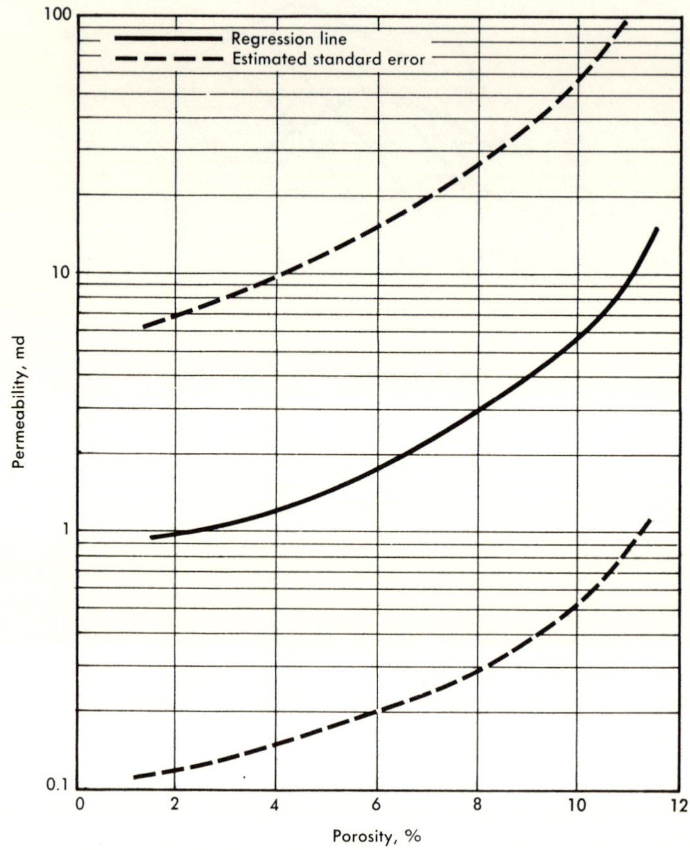

FIG. 4.6 Regression of permeability on porosity taken from FIG. 4.5.

analytical method would not. The analytical method is based on the assumption that the standard error will be the same at all levels of the independent variable.

Fig. 4.6 is the regression curve and its limits with the dependent variable, log K, transformed back to millidarcies. It can be seen that the probability limits are quite wide—the usual situation with permeability-porosity relationships in carbonates.

OIL INDUSTRY REFERENCES

Bulnes, A. C.: "An Application of Statistical Methods to Core Analysis of Dolomite Reservoirs," *Trans.,* AIME (1946) Vol. 165, p. 223.

Chatas, A. T. and Yankie, W. W., Jr.: "Application of Statistics to the Analysis of Production Decline Data," *Trans.,* AIME (1958) Vol. 213, p. 399.

Krumbein, W. C. and Monk, G. D.: "Permeability as a Function of the Size Parameters of Unconsolidated Sand," *Trans.,* AIME (1943) Vol. 151, p. 153.

McEwen, C. R.: "Material Balance Calculations with Water Influx in the Presence of Uncertainty in Pressures," *SPE Journal* (June, 1962) Vol. 2, p. 120.

Rachford, H. H., Jr. and Schultz, W. P.: "Some Useful Tables for Approximating Smooth Curves by Fifth-and-Lower Degree Polynomials," *Trans.,* AIME (1955) Vol. 204, p. 289.

Stevens, W. F. and Thodos, George: "Estimating Oil Reserves from Production Data," *The Petroleum Engineer* (February, 1961) Vol. 33, p. B46.

Van Everdingen A. F., Timmerman, E. H. and McMahon, J. J.: "Application of the Material Balance Equation to A Partial Water Drive Reservoir," *Trans.,* AIME (1953) Vol. 198, p. 51.

Walstrom, J. E., Brown, A. A. and Harvey, R. P.: "An Analysis of Uncertainty in Directional Surveying," AIME Trans. (1969), Vol. 246, p. I-515.

Willis, J. B.: "ABC's of Regression and Correlation Analysis," *Oil and Gas Journal* (January 23, 1961) Vol. 59, p. 97.

5

Multiple correlation and regression

Introduction

It has been shown in Chapter 4 on matrix applications that regressions involving more than one independent variable require a least squares analysis that simply is an extension of the principles used for a two-variable system. In fact, the computation used in multiple regression already has been demonstrated by the curvilinear regression problem of Chapter 4. The only difference is that the curvilinear regression takes the form

$$\hat{y} = b_o + b_1x + b_2x^2 + \cdots + b_px^p$$

whereas the multiple regression has the form

$$\hat{y} = b_o + b_1x_1 + b_2x_2 + \cdots + b_px_p$$

Actually, the two forms can be combined, for example

$$\hat{y} = b_o + b_1x_1 + b_2x_1^2 + b_3x_2, \text{ etc.}$$

or cross-products can be included, such as

$$\hat{y} = b_o + b_1x_1 + b_2x_2 + b_3x_1x_2, \text{ etc.}$$

The number of terms that can be used is restricted only by the patience of the investigator or the capacity of an electronic computer. The number of terms that *should* be used is another matter. One of the purposes of multiple correlation and regression analysis is to help the investigator decide which terms are significant and which are not.

Relationships between several variables can, like relationships between two variables, be described both by regression coefficients and by correlation coefficients; however, the purpose of a correlation analysis is quite different from that of a regression analysis, and it is best to keep the two types of problems separate.

If it is *known before* any data are analyzed that a relation exists between a set of variables, and the only purpose is to fit a regression equation to the data so that values of y can be predicted, then one should proceed with a regression analysis. But, if it *is not known* how or whether sets of variables are related, then it is best to set up a correlation matrix first before proceeding with a regression analysis.

First, an example of the procedure for direct solution of a multiple regression will be shown; then the procedure for solving the regression by use of correlation coefficients will be demonstrated.

Multiple regression

As an example of a regression problem, consider the data of Table 5A. Shown in the table are core analysis results from 20 samples of rock from a particular reservoir. The parameters measured on each sample were: connate water saturation (by the capillary-pressure method), porosity, and permeability.

The data from the 20 samples will be used to demonstrate the regression method. Since the connate (irreducible) water saturation depends upon the capillary properties of the rock, and these properties govern the porosity and permeability of the rock, it is desirable to have an expression for water saturation as a function of porosity and permeability, so that the saturation profile for a particular reservoir interval can be predicted from the permeability and porosity profiles of the interval.

As in previous problems a log transformation has been used to normalize the distribution of permeability. The products necessary for the Doolittle matrix solution of the regression

$$S_{wc} = b_o + b_1 \log_{10} K + b_2\phi$$

are found in Table 5A.

TABLE 5A

$y = S_{wc}$	$x_1 = \log_{10}K$	$x_2 = \phi$	x_1^2	x_2^2	x_1x_2	x_1y	x_2y
0.295	0.30	0.08	0.0900	0.0064	0.0240	0.08850	0.02360
0.320	1.02	0.21	1.0404	0.0441	0.2142	0.32640	0.06720
0.242	1.46	0.21	2.1316	0.0441	0.3066	0.35332	0.05082
0.140	1.50	0.12	2.2500	0.0144	0.1800	0.21000	0.01680
0.265	0.65	0.10	0.4225	0.0100	0.0650	0.17225	0.02650
0.335	0.71	0.16	0.5041	0.0256	0.1136	0.23785	0.05360
0.085	2.33	0.23	5.4289	0.0529	0.5359	0.19805	0.01955
0.170	1.80	0.21	3.2400	0.0441	0.3780	0.30600	0.03570
0.265	1.20	0.19	1.4400	0.0361	0.2280	0.31800	0.05035
0.292	0.91	0.15	0.8281	0.0225	0.1365	0.26572	0.04380
0.342	0.33	0.11	0.1089	0.0121	0.0363	0.11286	0.03762
0.095	1.80	0.14	3.2400	0.0196	0.2520	0.17100	0.01330
0.200	1.03	0.10	1.0609	0.0100	0.1030	0.20600	0.02000
0.200	1.31	0.14	1.7161	0.0196	0.1834	0.26200	0.02800
0.087	2.11	0.17	4.4521	0.0289	0.3587	0.18357	0.01479
0.145	1.22	0.09	1.4884	0.0081	0.1098	0.17690	0.01305
0.145	1.76	0.16	3.0976	0.0256	0.2816	0.25520	0.02320
0.165	1.56	0.20	2.4336	0.0400	0.3120	0.25740	0.03300
0.145	1.98	0.15	3.9204	0.0225	0.2970	0.28710	0.02175
0.070	2.50	0.20	6.2500	0.0400	0.5000	0.17500	0.01400
4.003	27.48	3.12	45.1436	0.5266	4.6156	4.56312	0.60663

$\Sigma y^2 = 0.9504$

The matrix of normal equations that will be used for the *abbreviated Doolittle solution* will have the form

			$g =$
N	Σx_1	Σx_2	Σy
	Σx_1^2	$\Sigma x_1 x_2$	$\Sigma x_1 y$
		Σx_2^2	$\Sigma x_2 y$

Table 5B is the forward Doolittle solution of the example problem. The solution proceeds in the fashion outlined in the section on matrix applications. (For a greater amount of procedural detail than shown below, refer to the curvilinear regression problem in Chapter 4).

TABLE 5B

					j =
(1)	20	27.48	3.12	4.003	54.603
(2)		45.1436	4.6156	4.56312	81.8023
(3)			0.5266	0.60663	8.86883
(4)	20	27.48	3.12	4.003	54.603
(5)	1	1.3740	0.1560	0.20015	2.73015
(6)		7.3861	0.3287	−0.9370	6.7778
(7)		1	0.044502	−0.12686	0.91764
(8)			0.02525	0.02386	0.04913
(9)			1	0.94495	1.945743

The various elements in Table 5B were obtained as follows:

Line 5: $20/20 = 1$; $27.48/20 = 1.3740$
$3.12/20 = 0.1560$; $4.003/20 = 0.20015$
$j_5 = 54.603/20 = 2.73015$

Line 6: $45.1436 - (27.48)(1.3740) = 7.3861$
$4.6156 - (27.48)(0.1560) = 0.3287$
$4.56312 - (27.48)(0.20015) = -0.9370$
$j_6 = 81.8023 - (27.48)(2.73015) = 6.7778$

Line 7: $7.3861/7.3861 = 1$; $0.3287/7.3861 = 0.044502$
$-0.9370/7.3861 = -0.12686$; $j_7 = 6.7778/7.3861 = 0.91764$

Line 8: $0.5266 - (3.12)(0.1560) - (0.3287)(0.044502) = 0.02525$
$0.60663 - (3.12)(0.20015) - (0.3287)(-0.12686) = 0.02386$
$j_8 = 8.86883 - (3.12)(2.73015) - (0.3287)(0.91764)$
$= 0.04913$

Line 9: $0.02525/0.02525 = 1$; $0.02386/0.02525 = 0.94495$
$j_9 = 0.04913/0.02525 = 1.945743$

The regression coefficients are produced by lines (5), (7) and (9) so that

$$b_2 = 0.94495$$
$$b_1 + 0.044502b_2 = -0.12686, \text{ and}$$
$$b_1 = -0.16891$$
$$b_0 + 1.3740b_1 + 0.1560b_2 = 0.20015, \text{ and}$$
$$b_0 = 0.28482$$

Therefore, the least squares regression of S_{wc} upon $\log_{10}$ K and ϕ is

$$\hat{S}_{wc} = 0.2848 - 0.1689 \log_{10} K + 0.9450\ \phi$$

An analysis of variance table now can be constructed and F tests can be made to verify the significance of the regression coefficients b_1 and b_2. This is presented in Table 5C.

The F-test shows that b_1 and b_2 both are highly significant, i.e. there is less than 1 chance in 100 of either coefficients being, in fact, insignificant. The coefficient b_1 refers to $\log_{10}$ K and the coefficient b_2 refers to ϕ. What Table 5C tells us, in terms of the original data, is that both permeability and porosity have a highly significant effect upon connate water saturation.

TABLE 5C

Source of variance	Sum of squares	Degrees of freedom	Mean square	F
Σy^2	0.950411	20		
b_0	0.801200	1		
Mean	0.149211	19		
b_1	0.118868	1	0.118868	$F_{1,\,17} = 259$**
b_2	0.022547	1	0.022547	$F_{1,\,17} = 49$**
Residual	0.007796	17	0.000459	

This completes the information that can be gotten from the forward portion of the Doolittle solution. To set confidence limits upon the estimates of y_i, b_1 and b_2, it is necessary to perform the backward portion of the Doolittle solution, thus completing the matrix inversion in the manner following.

$$c_{22} = 1/A_{22} = 1/0.02525 = 39.6039$$

$$c_{12} = -B_{12}c_{22} = (-0.044502)(39.6039) = -1.76245$$

$$c_{02} = -B_{01}c_{12} - B_{02}c_{22} = (-1.3740)(-1.76245) - (0.1560)(39.6039) = -3.75660$$

$$c_{11} = 1/A_{11} - B_{12}c_{12} = 1/7.3861 - (0.044502)(-1.76245) = 0.213823$$

$$c_{01} = -B_{01}c_{11} - B_{02}c_{12} = (-1.3740)(0.213823) - (0.1560)(-1.76245) = -0.018851$$

$$c_{00} = 1/A_{00} - B_{01}c_{01} - B_{02}c_{02} = 1/20 - (1.3740)(-0.018851) - (0.1560)(-3.75660) = 0.66193$$

As a check on the inversion computations

$$a_{02}c_{02} + a_{12}c_{12} + a_{22}c_{22} = (3.12)(-3.75660) + (4.6156)(-1.76245) + (0.5266)(39.6039) = 1.00006$$

$$a_{01}c_{01} + a_{11}c_{11} + a_{21}c_{21} = (27.48)(-0.018851) + (45.1436)(0.2138) + (4.6156)(-1.76245) = 0.9989$$

$$a_{00}c_{00} + a_{10}c_{10} + a_{20}c_{20} = (20)(0.66193) + (27.48)(-0.018851) + (3.12)(-3.7566) = 0.99998$$

The variances on the estimated regression coefficients are

$$s^2(b_1) = s_e^2c_{11} = (0.000459)(0.2138) = 0.000098$$
$$s^2(b_2) = s_e^2c_{22} = (0.000459)(39.60) = 0.01818$$
$$\text{Cov}\ (b_1b_2) = s_e^2c_{12} = (0.000459)(-1.7625) = -0.000809$$

and the 95% confidence limits on the estimates of b_1 and b_2 at 17 degrees of freedom are

$$b_1 \pm t_{1/2a}\sqrt{s_e^2c_{11}} = -0.1689 \pm 2.110\sqrt{0.98 \times 10^{-4}} = -0.1898 \text{ and } -0.1480$$

and

$$b_2 \pm t_{1/2a}\sqrt{s_e^2c_{22}} = 0.9450 \pm 2.110\sqrt{0.01818} = 0.6605 \text{ and } 1.2295$$

The standard errors of the regression coefficients, and the confidence limits calculated therefrom, measure the overall uncertainty of each estimated regression coefficient taken separately. The uncertainty as to the true value of (for instance) b_1 springs in part, however, from uncertainty about the true values of the other coefficients; the estimate b_1 is obtained after making allowance for the apparent effects of the other factors.

Any errors in estimation of b_2, b_3, etc. in any regression will cause related errors in b_1, so that the separate estimates will be correlated with each other. The correlation coefficient measuring the correlation between b_i and b_j in any multiple regression is

$$r(b_ib_j) = c_{ij}/\sqrt{c_{ii}c_{jj}} \qquad (5.1)$$

In the example problem

$$r(b_1b_2) = c_{12}/\sqrt{c_{11}c_{22}} = -1.7624/\sqrt{0.2138 \times 39.6039} = -0.606^{**}$$

This correlation coefficient is associated with (20–3) = 17 degrees of freedom and is highly significant.

The negative correlation between b_1 and b_2 implies that errors of estimation which will cause b_1 to be high will tend also to cause the estimate of b_2 to be low, and conversely. The correlation is not perfect, so there will not be perfect association between estimates of b_1 and the corresponding estimates of b_2. It is high enough, however, for the scatter in the values of b_1 associated with any particular value of b_2 to be appreciably less than the overall scatter of possible values of b_1.

Thus, confidence limits for b_1 at any specified value of b_2 will be narrower than those for b_2 unspecified. Because of the association, also, the positions of the limits will change as the value of b_2 is changed. It can, therefore, be misleading to consider b_1 without also considering b_2.

Most statistical texts give a method for placing confidence limits on simultaneous estimates of b_i and b_j, but these are of limited interest in most engineering problems.

The regression coefficients are available from the inverse matrix and may be recalculated as a final check on the matrix inversion. The b's are found by multiplying the elements of each row of the C matrix by the elements of the G matrix (Table 5B) in the manner,

$$b_0 = c_{00}g_0 + c_{01}g_1 + c_{02}g_2$$
$$b_1 = c_{01}g_0 + c_{11}g_1 + c_{12}g_2,$$

etc. In the example problem

$$b_0 = (0.6619)(4.003) - (0.01885)(4.5631) - (3.757)(0.6066) = 0.2846$$
$$b_1 = (-0.01885)(4.003) + (0.2138)(4.5631) - (1.7624)(0.6066) = -0.1689$$
$$b_2 = (-3.7566)(4.003) - (1.7624)(4.5631) + (39.6039)(0.6066) = 0.9440$$

which checks reasonably well the values calculated from the forward Doolittle solution.

The engineer's main interest probably will center around the question "how good is a value of y calculated from the regression equation?"

The variance of $\hat{y}$ is given by Equation 4.20, which, for this problem, can be written

$$s^2(\hat{S}_{wc}) = s_e^2/N + s^2(b_1)(\log K_i - \overline{\log K})^2 + s^2(b_2)(\phi_i - \overline{\phi})^2 + 2\,\mathrm{cov}(b_1b_2)(\log K_i - \overline{\log K})(\phi_i - \overline{\phi})$$

To illustrate, if $\log_{10} K = 1.02$ and $\phi = 0.21$

$$\hat{S}_{wc} = 0.2848 - (0.1689)(1.02) + (0.9450)(0.21) = 0.311$$

and

$$s^2(\hat{S}_{wc}) = (0.000459/20) + (0.000098)(1.02 - 1.374)^2 + (0.01818)(0.21 - 0.156)^2 + (2)(-0.000809)(1.02 - 1.374)(0.21 - 0.156) = .000119$$

The 95% confidence limits on $\hat{S}_{wc}$ are

$$0.311 \pm 2.110\ \sqrt{1.19 \times 10^{-4}} = 0.288 \text{ and } 0.334$$

The variance of a predicted $\hat{S}_{wc}$ is

$$s^2(\hat{S}_{wc}) + s_e^2 = 0.000119 + 0.000459 = 0.000578$$

and the 95% confidence limits on a prediction of $\hat{S}_{wc}$ are

$$0.311 \pm 2.110\ \sqrt{5.78 \times 10^{-4}} = 0.260 \text{ and } 0.362$$

On use of significance tests it must be appreciated that nonsignificance of a particular regression coefficient does not in any way imply that the independent variable concerned does not affect, or is not related to, the dependent variable. It implies merely that, at the level of significance adopted, the confidence limits for the estimated effect, or slope, included zero as a possible value. Where, additionally, the confidence limits are so narrow that any true effect which may exist is small and of little importance over the range for which the relationship is required, there may be justification for ignoring the term.

In other cases, however, one may be justified only in concluding that further work is required to define the relationship more precisely. The uncritical use of tests of significance should be scrupulously avoided. Likewise the analysis of variance should not be applied indiscriminately.

To illustrate, if some of the fixed variables are not truly independent —as is the case with permeability and porosity in the previous problem —contradictory conclusions sometimes can be drawn from the different tests of significance. For example, it is possible that the F-tests may show that the reduction of variance due to a particular variable is significant, but at the same time the t-test on the regression coefficient for this variable may show it to be insignificant. This can happen whenever b_i and b_j are highly correlated so that the actual influence of x_i on y is split into a part contributed by b_i and another part contributed by b_j, making it impossible to interpret b_i as the change in y when x_i varies while x_j is held constant because, in fact, x_j varies when x_i varies.

This hypothetical situation points up some of the difficulties of interpreting regression analyses when some of the fixed variables are closely related to one another. If two fixed variables are highly correlated, it is unrealistic to assume that one can be held constant while the other varies. In some analyses, it may be impractical to consider that *any* of the fixed variables can be held constant while any one is varied. *In this case the regression equation should be considered as an entity, and significance tests of individual components should not be given too much credence.*

In summary the investigator should know something about the environment he is studying beyond what statistics alone will tell him.

Multiple correlation

In the discussion of regressions the importance of intercorrelations between fixed variables are emphasized. The engineer may encounter some problems, such as those involving rock parameters, where there is little theoretical or experimental basis for postulating the relationships between parameters. He will simply have to measure parameters which he thinks are significant and try them in an analysis.

One approach he could use would be to do a regression analysis on all the variables, then do another with one variable dropped, and observe the effect on $s(\hat{y})$. Those variables which did not contribute to a significant reduction of $s(\hat{y})$ could then be eliminated. A less tedious method, however, is multiple correlation analysis.

The correlation coefficient has been defined in Chapter 4 as a measure of the linear relationship between two variables, and the mathematical expression for it is given by Equation 4.2. This is called a *simple correlation.* There are other correlations that enter into a correlation analysis that should be defined.

Primary correlations are simple correlations between the dependent variable and one of the fixed variables. The coefficient is expressed symbolically as r_{oi}.

Interfactor correlations are simple correlations between two of the fixed variables. Their coefficient is expressed as r_{ij}.

A *partial correlation* is a correlation between the dependent and one or more of the fixed variables, with the effect of the other fixed variables removed. The partial correlation coefficient is designated $r_{oi \cdot jk}$ which would be read as "the correlation coefficient of y and x_i with x_j and x_k removed. This means that x_j and x_k are held constant while the correlation between y and x_i is measured.

Partial correlation coefficients are of limited interest in most engineering problems, but they can be useful in helping the engineer decide whether one or more of the independent variables can be dropped without damaging the overall correlation. This application will be illustrated shortly.

Each partial correlation coefficient is related to a reduction of some particular sum of squares. For example $r_{30 \cdot 12}$, the partial coefficient between y and x_3 with x_1 and x_2 fixed, may be defined as [residual ss of y after removing regression on x_1, x_2 and x_3] ÷ [residual ss of y after removing regression on x_1 and x_2] $= 1 - r^2_{30 \cdot 12}$. A more practical way to express the partial coefficients is in terms of the simple correlation coefficients. Two representative expressions are:

$$r_{01 \cdot 2} = (r_{01} - r_{02} r_{12}) / [(1 - r_{02}^2)(1 - r_{12}^2)]^{1/2} \tag{5.2}$$

and

$$r_{01 \cdot 234} = (r_{01 \cdot 23} - r_{04 \cdot 23} r_{14 \cdot 23}) / [(1 - r_{04 \cdot 23}^2)(1 - r_{14 \cdot 23}^2)]^{1/2} \tag{5.3}$$

The *multiple correlation coefficient,* denoted by R, measures the *overall extent of association* between values of the dependent variable and values of the fixed variables. There are several ways to compute R. If the analysis of variance table is used,

$$\begin{aligned} R^2 = &\ (\text{Total sum of squares due to } b_1, b_2, b_3, \text{etc.}) \\ &\ \div (\text{Total sum of squares about the mean}) \end{aligned} \tag{5.4}$$

It was stated previously in the discussion of "Linear Relationships between Variables" that the square of the correlation coefficient is the total variance accounted for by correlation. It follows, then, that the fraction of the variance that is residual after correlation is $1 - R^2$. This leads to the following expression for residual sum of squares:

$$\text{residual sum of squares} = (1 - R^2)[\Sigma y^2 - (\Sigma y)^2/N] \qquad (5.5)$$

where $\Sigma y^2 - (\Sigma y)^2/N$ is the sum of squares due to the mean.

The multiple correlation coefficient is simply a partial correlation coefficient with the effects of all the independent variables removed, therefore for the particular problem in question $R^2 = r_{0.12}^2$, which can be expressed in terms of the covariances as

$$\begin{aligned} r_{0.12}^2 &= [b_1 \operatorname{cov}(x_1 y) + b_2 \operatorname{cov}(x_2 y)]/s_y^2 \\ &= [b_1(\Sigma x_1 y - \Sigma x_1 \Sigma y/N) + b_2(\Sigma x_2 y - \Sigma x_2 \Sigma y/N)] \\ &\qquad [\Sigma y^2 - (\Sigma y)^2/N] \end{aligned} \qquad (5.6)$$

This expression can be expanded to include as many independent variables as needed simply by adding all the products $b_i(\Sigma x_i y - \Sigma x_i \Sigma y/N)$ into the numerator.

Still another approach to the multiple coefficient is through β-factors. These will be discussed later in connection with correlation matrices.

The primary correlations from the connate water-porosity-permeability data of the foregoing regression problem given by Equation 4.2 are

$$\begin{aligned} r_{01} &= [4.5631 - (27.48)(4.003)/20] \\ &\quad \div \{[0.9504 - (4.003)^2/20][45.1436 - (27.48)^2/20]\}^{1/2} \\ &= -0.892 \\ r_{01}^2 &= 0.796^{**} \end{aligned}$$

and

$$\begin{aligned} r_{02} &= [0.6066 - (4.003)(3.12)/20] \\ &\quad \div \{[0.9504 - (4.003)^2/20][0.5266 - (3.12)^2/20]\}^{1/2} \\ &= -0.232 \\ r_{02}^2 &= 0.054 \end{aligned}$$

The interfactor correlation coefficient is

$$\begin{aligned} r_{12} &= [4.6156 - (27.48)(3.12)/20] \\ &\quad \div \{[45.1436 - (27.48)^2/20][0.5266 - (3.12)^2/20]\}^{1/2} \\ &= 0.605 \\ r_{12}^2 &= 0.367^{**} \end{aligned}$$

The coefficient, r_{01}, measures the correlation between connate water saturation and the log of permeability; r_{02} measures the correlation between connate water saturation and porosity, and r_{12} measures the correlation between permeability and porosity. Checking these coefficients at $(20 - 3) = 17$ degrees of freedom in the left-hand side of Table A6 reveals that r_{01}^2 and r_{12}^2 are significant at the 99% level, but r_{02}^2 does not reach significance. Thus, there is a high negative correlation between saturation and permeability and a high positive correlation between porosity and permeability, but the correlation between porosity and saturation is not high enough to be significant at the customary 95% confidence level. (More extensive tables than A6 can be found in books of statistical tables. These would permit one to determine the confidence level at which r_{02}^2 is significant).

The fact that r_{02}^2 is not significant at the 95% confidence level does not of itself mean that porosity shouldn't be included in the correlation, because the overall effect of the porosity-permeability correlation is unknown at this point. This will be investigated by way of the partial correlation coefficient, $r_{01.2}$.

The sums of squares from Table 5C can be substituted in Equation 5.4 to get the multiple correlation coefficient,

$$R^2 = (0.1189 + 0.0225)/0.1492 = 0.948^{**}$$

The partial correlations, given by Equation 5.2 are

$$\begin{aligned} r_{01\cdot 2} &= [-0.8923 - (-0.232)(0.605)]/\{[1 - (0.232)^2][1 - (0.605)^2]\}^{1/2} \\ &= -0.971 \\ r_{01\cdot 2}^2 &= 0.942^{**} \end{aligned}$$

and

$$\begin{aligned} r_{02\cdot 1} &= [-0.232 - (-0.892)(0.605)]/\{[1 - (-0.892)^2][1 - (0.605)^2]\}^{1/2} \\ &= 0.855 \\ r_{02\cdot 1}^2 &= 0.731^{**} \end{aligned}$$

These partial correlations reveal that the overall correlation would hardly be affected if it had been assumed that porosity is constant (because $r_{01\cdot 2}^2$ is practically as large as R^2); but the correlation would be considerably reduced if permeability were to be assumed constant and only porosity allowed to vary.

A convenient method of treating correlations is by use of β-factors.

These factors are especially useful in problems involving a large number of variables, because they are easily arranged in matrix form.

A β-factor is a partial regression coefficient expressed in transformed units so that the coefficient is independent of the original units of measurement. The transformation is accomplished by dividing all observations of each variable by the standard deviation of that series of observations. Beta is defined mathematically as

$$\beta_1 = b_1 s(x_1)/s(y),\ \beta_2 = b_2 s(x_2)/s(y), \text{ etc.} \tag{5.7}$$

The β's are related to the regression equation as follows:

when β's are substituted in the regression equation

$$y = b_0 + b_1 x_1 + b_2 x_2$$

the new expression is

$$y = b_0 + \beta_1 s(y)[x_1/s(x_1)] + \beta_2 s(y)[x_2/s(x_2)]$$

also, (see Equation 4.8),

$$\overline{Y} = b_0 + \beta_1 s(y)[\overline{X}_1/s(x_1)] + \beta_2 s(y)[\overline{X}_2/s(x_2)]$$

Simultaneous solution of the preceding two equations yields

$$(y - \overline{Y}) = \beta_1 s(y)[(x_1 - \overline{X}_1)/s(x_1)] + \beta_2 s(y)[(x_2 - \overline{X}_2)/s(x_2)]$$

dividing through by s(y),

$$(y - \overline{Y})/s(y) = \beta_1[(x_1 - \overline{X}_1)/s(x_1)] + \beta_2[(x_2 - \overline{X}_2)/s(x_2)]$$

For simplification, designate

$$(y - \overline{Y})/s(y) = y';(x_1 - \overline{X}_1)/s(x_1) = x_1';(x_2 - \overline{X}_2)/s(x_2) = x_2'$$

The generalized multiple regression equation would be

$$y' = \beta_0 + \beta_1 x_1' + \beta_2 x_2' + \cdots + \beta_p x_p' \tag{5.8}$$

A system of normal equations will be

$$\beta_0 \Sigma x_0'^2 + \beta_1 \Sigma x_0' x_1' + \beta_2 \Sigma x_0' x_2' + \cdots + \beta_p \Sigma x_0' x_p' = \Sigma x_0' y'$$
$$\beta_0 \Sigma x_1' x_0' + \beta_1 \Sigma x_1'^2 + \beta_2 \Sigma x_1' x_2' + \cdots + \beta_p \Sigma x_1' x_p' = \Sigma x_1' y'$$
$$\cdots\cdots\cdots\cdots\cdots\cdots\cdots\cdots\cdots\cdots\cdots\cdots\cdots\cdots\cdots\cdots$$
$$\beta_0 \Sigma x_p' x_0' + \beta_1 \Sigma x_p' x_1 + \beta_2 \Sigma x_p' x_2' + \cdots + \beta_p \Sigma x_p'^2 = \Sigma x_p' y'$$

(See Appendix B for development of the matrix of normal equations.)

All the β_0 terms in this set are equal to zero, and

$$\Sigma x_i' x_j' = r_{ij}$$

so the normal equations will reduce to

$$\begin{aligned} r_{11}\beta_1 + r_{12}\beta_2 + r_{13}\beta_3 + \cdots + r_{1p}\beta_p &= r_{01} \\ r_{21}\beta_1 + r_{22}\beta_2 + r_{23}\beta_3 + \cdots + r_{2p}\beta_p &= r_{02} \\ \cdots\cdots\cdots\cdots\cdots\cdots\cdots\cdots\cdots\cdots \\ r_{p1}\beta_1 + r_{p2}\beta_2 + r_{p3}\beta_3 + \cdots + r_{pp}\beta_p &= r_{0p} \end{aligned}$$

The r's are simple correlation coefficients which can be calculated by application of Equation 4.2. The subscripts on the r's indicate the variables to which the particular correlation coefficient applies. After the r-values have been filled in, the β's can be solved by the Forward Doolittle method. The regression coefficients (b_1, b_2, b_3, etc.) can then be calculated by use of Equation 5.7. The s(y) term of Equation 5.7 is the standard deviation of the measured y-values, $s(x_1)$ is the standard deviation of x_1, etc. These standard deviations are computed by use of Equation 1.11. After the regression coefficients have been computed, b_0 can be calculated by substitution of numerical values into the regression equation, thus completing the regression equation.

The multiple correlation coefficient is

$$R^2 = r_{01}\beta_1 + r_{02}\beta_2 + r_{03}\beta_3 + \cdots + r_{0p}\beta_p \tag{5.9}$$

and the confidence limits on a predicted value of y are

$$y_i \pm t_{1/2\alpha} s(y) \sqrt{(1 - R^2)(N - 1)/(N - p - 1)} \tag{5.10}$$

A suggested procedure for multiple correlation problems is:

(1) Compute the primary correlation coefficients.

(2) Inspect the primary correlation coefficients and discard from further consideration those variables which are not significant at a low level, say 0.1 probability. This procedure discards the possibility of independent secondary effects, so if the investigator suspects such an effect, he may check the interfactor correlations then test the significance of the primary correlation using the products of the interrelated variables before deciding whether to discard them.

(3) Set up the normal equations in r and β and compute the β's by the Forward Doolittle method.

(4) Compute $r_{01}\beta_1$, $r_{02}\beta_2$, etc.—each of these products equals the fraction of the variance of the fixed variable that is linearly related to the other variables—and discard from further calculations those variables whose $r_{0i}\beta_i$ is either very small or negative. It will be found at times that a small correlation between a fixed variable and the dependent variable will be more than accounted for by interfactor correlations, causing $r_{0i}\beta_i$ for that factor to be negative. This is interpreted as zero relationship.

(5) Recompute the β's for the normal equations, using only those variables which have survived the elimination of step (4). If additional variables become negligible on this second computation, they should be eliminated and the β's again recomputed.

(6) Compute the multiple correlation coefficient. If it is judged significant, complete the regression equation.

(7) If a curvilinear relationship is suspected between the dependent variable and one of the fixed variables, plot the values of the dependent variable (corrected to the average value of the variables which should be held constant) against the values of the variable in question. If a curvilinear function is warranted, repeat the correlation calculation, using a transformation or a higher order polynomial to represent that variable.

(8) When the regression equation is satisfactory, compute $s(\hat{y})$, using Equation 4.20.

Design of Experiments

Most of the time the petroleum engineer is confronted with experimental data that were taken in uncontrolled or partially controlled tests. If, however, he ever should get an opportunity to control an experiment, he should be aware of the advantages of designed experimentation.

Experimental design is a large and complex field, most of which is beyond the scope of this resumé, but some of the underlying principles and several of the simpler designs for multiple and curvilinear regression experiments will be given here.

Of the coefficients in the least squares equations, the sums of squares are essentially positive and, in practice, different from zero; for if a sum of squares, say $\Sigma(x_1 - \overline{X}_1)^2$, were equal to zero, this would imply that all observed values of x_1 were the same hence that $\Sigma(x_1 - \overline{X}_1)(x_2 - \overline{X}_2)$, $\Sigma(x_1 - \overline{X}_1)(x_3 - \overline{X}_3)$, etc. were also zero. Thus, all coefficients of b_1 in the least squares equations would be zero and b_1 would be completely indeterminate; whatever value were assigned to it, the equations would still be satisfied. In other words, if x_1 does not vary in the set of observations, no information can be obtained about the effects of variations in x_1.

It is possible, however, to have conditions that will permit the sum of all cross-products containing x_1 to be zero while still permitting variations of x_1. The vanishing of a sum of products results in what is known as *orthogonality*. Orthogonality could occur by chance, but this is highly unlikely. When, however, the values of x may be assigned at will, as in a planned investigation rather than an examination of accumulated data, orthogonality may be deliberately imposed and will lead to a considerable saving of effort in the analysis of the results, as well as to some gain in the precision of estimation. The principle underlies the whole development of statistical design of experiments and is most clearly evident in the balanced factorial designs for regression problems.

It should be noted that the vanishing of the product terms does not necessarily imply that the "independent," or fixed, variables are not related. In fact, orthogonality can be applied to curvilinear regressions where the relation between x, x^2, x^3, etc. is obvious. The mutual independence of the so-called "independent" variables is of little importance in the theoretical considerations of orthogonal systems.

When the sums of products vanish, the least squares equations reduce to

$$b_k\Sigma(x_k - \overline{X}_k)^2 = \Sigma(x_k - \overline{X}_k)(y - \overline{Y}) \qquad (5.11)$$

where

$$k = 1, p$$

All covariances are zero, so it does not matter whether the x-variables are considered separately or together. Each regression coefficient thus reflects only the effect of the variable concerned, implying that each may be tested separately for significance, and the effect of adding or suppressing a term may be ascertained at once.

The expression for variance of a regression coefficient in an orthogonal system is

$$s^2(b_k) = s_e^2/\Sigma(x_k + \overline{X}_k)^2 \tag{5.12}$$

Only the diagonal elements of the A matrix will exist. For example, the matrix

$$\begin{array}{ccc|c} N & \Sigma x_1 & \Sigma x_2 & \Sigma y \\ & \Sigma x_1^2 & \Sigma x_1 x_2 & \Sigma x_1 y \\ & & \Sigma x_2^2 & \Sigma x_2 y \end{array}$$

will reduce to

$$\begin{array}{ccc|c} N & 0 & 0 & \Sigma y \\ & \Sigma x_1^2 & 0 & \Sigma x_1 y \\ & & \Sigma x_2^2 & \Sigma x_2 y \end{array}$$

The purpose of factorial designs is to order the elements of an experiment so that orthogonality will result. Factorial designs for large numbers of "independent" variables are possible but, of course, they grow more complex as the number of variables increase. They are known as factorial designs because there will have to be P^q sets of measurements to get an orthogonal design where there are q variables observed at P levels.

For an illustration of a factorial design at two levels consider a hypothetical problem where some variable, y, is read at two different temperatures and times. Since temperature and time both can be controlled, it is possible to design an orthogonal experiment. The minimum requirement is that $2^2 = 4$ observations be made and that a value of y be read at each combination of temperature and time, as follows.

y	x_1 = time, hr	x_2 = temperature, °F.
1	1	240
6	5	240
7	1	280
18	5	280

Coding must be done so that all x-values become either +1 or −1, therefore

$$x_1' = (x_1 - 3)/2 \text{ and}$$
$$x_2' = (x_2 - 260)/20$$

causing the coded data to become

y	x_1'	x_2'
1	−1	−1
6	+1	−1
7	−1	+1
18	+1	+1

so the forward Doolittle solution will be

4	0	0	32
	4	0	16
		4	18
4	0	0	32
1	0	0	8
	4	0	16
	1	0	4
		4	18
		1	4.5

making the regression equation

$$\hat{y} = 8 + 4.0x_1' + 4.5x_2'$$

The analysis of variance is shown in Table 5D.

TABLE 5D

Source of variance	Sum of squares	Degrees of freedom	Mean square	F
Σy^2	410	4		
b_0	256	1		
Mean	154	3		
b_1	64	1	64	$F_{1,1} = 7.1$
b_2	81	1	81	$F_{1,1} = 9.0$
Residual	9	1	9	

The non-significance of b_1 and b_2 indicates a high residual variance and a large lack of fit, so it is obvious that either an important variable has been left out or the experimental error is quite large. This question can be resolved by replication of measurements, that is, taking repeat measurements of y at each level of x_1 and x_2. Suppose the set of data after replication is

y	x_1'	x_2'
1.0	−1	−1
1.7	−1	−1
6.0	+1	−1
5.2	+1	−1
7.0	−1	+1
7.9	−1	+1
18.0	+1	+1
19.2	+1	+1

The forward Doolittle solution is

8	0	0	66.0
	8	0	30.8
		8	38.2
8	0	0	66.0
1	0	0	8.25
	8	0	30.800
	1	0	3.85
		8	38.20
		1	4.775

so

$$\hat{y} = 8.25 + 3.85x_1' + 4.775x_2'$$

If N_r is the number of replicate sets and k is the number of repeat measurements in each set, the sum of squares due to measurement error is

$$\text{error s.s.} = \sum_{j=1}^{N_r} [\Sigma x_{ij}^2 - (\Sigma x_{ij})^2/k] \tag{5.13}$$

$$= [3.89 - (2.7)^2/2] + [63.04 - (11.2)^2/2] + [111.41 - (14.9)^2/2] + [692.64 - (37.2)^2/2] = 1.69$$

Table 5E is the new analysis of variance.

TABLE 5E

Source of variance	Sum of squares	Degrees of freedom	Mean square	F
Σy^2	870.980	8		
b_0	544.500	1		
Mean	326.480	7		
b_1	118.580	1	118.6	$F_{1,4} = 282^{**}$
b_2	182.405	1	182.4	$F_{1,4} = 434^{**}$
Residual	25.495	5	5.1	
Lack of fit	23.805	1	23.8	$F_{1,4} = 57^{**}$
Experimental error	1.690	4	0.42	

The F-test based on experimental error shows b_1, b_2 and lack of fit all to be highly significant, therefore one should look for (1) additional variables, (2) a nonlinear form of x_1 or x_2, or (3) the product x_1x_2 in seeking ways to improve the fit.

If the problem were expanded to include a third variable, say pressure, $2^3 = 8$ sets of observations would have to be made, and the table of data (without replications) might appear as

y	$x_1 =$ Time, hr	$x_2 =$ Temperature, °F.	$x_3 =$ Pressure, psia
1.0	1	240	2,000
6.0	5	240	2,000
2.5	1	240	5,000
7.2	5	240	5,000
7.0	1	280	2,000
18.0	5	280	2,000
8.3	1	280	5,000
23.0	5	280	5,000

The advantages of orthogonality in the fitting of multiple regressions are impressive and it is natural to seek a similar means of simplifying the curvilinear regression. It is not possible in general to arrange for orthogonality of successive powers of a series of x-values. In fitting the regression, however, x^2 is treated exactly as if it were a second independent variable in a multiple regression, therefore

$$y = b_0 + b_1x + b_2x^2 + \cdots$$

can be written

$$y = A + B(b_1 + a_1x) + C(b_2 + a_2x + a_2x^2) + \cdots$$

or

$$y = A + B\,\xi_1 + C\,\xi_2 + \cdots$$

where ξ_k is the deviation of x^k from a multiple linear regression of x^k on all powers of x up to x^{k-1}. For example, the second order polynomial, ξ_2, is the residual deviation of x^2 from a linear regression of x^2 on x. The ξ-functions can be chosen so as to be mutually orthogonal, and the coefficients of a curvilinear regression equation written in terms of orthogonal ξ-functions are given simply by

$$b_k = \sum_{i=1}^{N} y_i(\xi_k)_i / \sum_{i=1}^{N}(\xi_k)_i^2 \tag{5.14}$$

The standard error of estimate of b_k will be

$$s(b_k) = s_e / \sqrt{\sum_{i=1}^{N}(\xi_k)_i^2} \tag{5.15}$$

Provided ξ-functions are available, successive terms may be added to the regression equation without difficulty and their significance tested at once; because of orthogonality the coefficients and their contributions to the total sum of squares are unaltered by the addition or deletion of other terms.

While use of orthogonal functions is theoretically possible for any set of x-values, the construction and use of them would be extremely tedious except when the experiment has been designed so that the x's are equally spaced over the whole range covered. For this special case, ξ-functions and the corresponding sums of squares have been tabulated. For values of x spaced at unit interval the first four ξ-functions are given by

$$\xi_1 = x - \overline{X} \cdots \tag{5.16}$$
$$\xi_2 = (x - \overline{X})^2 - (N^2 - 1)/12 \cdots \tag{5.17}$$

$$\xi_3 = (x - \overline{X})^3 - \left(\frac{3N^2 - 7}{20}\right)(x - \overline{X}) \cdots \tag{5.18}$$

$$\xi_4 = (x - \overline{X})^4 - \left(\frac{3N^2 - 13}{14}\right)(x - \overline{X})^2 + 3\,\frac{(N^2 - 9)(N^2 - 1)}{560} \tag{5.19}$$

where N is the number of observations.

Suppose values of x are 1, 2, 3, 4, and 5. Then, the corresponding values of ξ_3 are $-6/5$, $+12/5$, 0, $-12/5$ and $+6/5$. These numbers can be made integral by multiplying by 5/6 so that

$$\xi_3' = 5/6\ \xi_3$$

Such prime values of ξ can be used conveniently to fit curvilinear regressions because they are also orthogonal functions of x, so equations 5.14 and 5.15 can be written

$$b_k = \sum_{i=1}^{N} y_i(\xi_k')_i / \sum_{i=1}^{N} (\xi_k')_i^2 \tag{5.20}$$

$$s(b_k) = s_e / \sqrt{\sum_{i=1}^{N} (\xi_k')_i^2} \tag{5.21}$$

The contribution of b_k to the total sum of squares will be

$$\left[\sum_{i=1}^{N} y_i(\xi_k')_i\right]^2 / \sum_{i=1}^{N} (\xi_k')_i^2 \tag{5.22}$$

Tables giving ξ_1 to ξ_5 for various values of N can be found in *Statistical Tables for Biological, Agricultural and Medical Research,* by R. A. Fisher and F. Yates, Oliver and Boyd (London, 1953).

The tabular entries generally are ξ' rather than actual ξ-values.

Of course, the Doolittle method can be used just as in any multiple regression. The orthogonal Doolittle matrix for a third-order polynomial would be

N	0	0	0	Σy
	$\Sigma\xi_1^2$	0	0	$\Sigma\xi_1 y$
		$\Sigma\xi_2^2$	0	$\Sigma\xi_2 y$
			$\Sigma\xi_3^2$	$\Sigma\xi_3 y$

OIL INDUSTRY REFERENCES

Arps, J. J., Brons, Folkert and Van Everdingen, A. F.: "A Statistical Study of Recovery Efficiency," *Bull.,* API (Oct., 1967) D14.

Bennion, D. W. and Griffiths, J. C.: "A Stochastic Model for Predicting Variations in Reservoir Rock Properties," Trans. AIME (1966), Vol. 237, p. II-9.

Guthrie, R. K. and Greenberger, Martin H.: "The Use of Multiple Correlation Analysis for Interpreting Petroleum Engineering Data," API, Division of Production (1955).

Khan, A. M.: "An Empirical Approach to Waterflood Predictions," *Journal of Petroleum Technology* (May, 1971) p. 565.

STATISTICAL REFERENCES

Alger, Phillip L.: Mathematics for Science and Engineering, McGraw-Hill, New York (1957).

Anderson, R. L. and Bancroft, T. A.: Statistical Theory in Research, McGraw-Hill, New York (1952).

Brownlee, K. A.: Statistical Theory and Methodology in Science and Engineering, John Wiley and Sons, New York (1960).

Davies, O. L.: Statistical Methods in Production and Research, Oliver and Boyd, London and Edinburgh (1957).

Dixon, W. J. and Massey, F. J., Jr.: Introduction to Statistical Analysis, McGraw-Hill, New York (1957).

Gore, W. L.: Statistical Methods for Chemical Experimentation, Interscience Publishers, New York (1952).

Snedecor, George W.: Statistical Methods, The Iowa State College Press, Ames, Iowa (1940).

Villars, D. S.: Statistical Design and Analysis of Experiments for Development Research, Wm. C. Brown, Co., Dubuque, Iowa (1951).

Volk, William: "Industrial Statistics" Chemical Engineering (March, 1956).

Worthing, A. G. and Geffner, Joseph: Treatment of Experimental Data, John Wiley and Sons, New York (1943).

APPENDIX **A**

Statistical tables

TABLE A-1
PERCENTILES OF THE t DISTRIBUTION

df	$t_{.60}$	$t_{.70}$	$t_{.80}$	$t_{.90}$	$t_{.95}$	$t_{.975}$	$t_{.99}$	$t_{.995}$
1	.325	.727	1.376	3.078	6.314	12.706	31.821	63.657
2	.289	.617	1.061	1.886	2.920	4.303	6.965	9.925
3	.277	.584	.978	1.638	2.353	3.182	4.541	5.841
4	.271	.569	.941	1.533	2.132	2.776	3.747	4.604
5	.267	.559	.920	1.476	2.015	2.571	3.365	4.032
6	.265	.553	.906	1.440	1.943	2.447	3.143	3.707
7	.263	.549	.896	1.415	1.895	2.365	2.998	3.499
8	.262	.546	.889	1.397	1.860	2.306	2.896	3.355
9	.261	.543	.883	1.383	1.833	2.262	2.821	3.250
10	.260	.542	.879	1.372	1.812	2.228	2.764	3.169
11	.260	.540	.876	1.363	1.796	2.201	2.718	3.106
12	.259	.539	.873	1.356	1.782	2.179	2.681	3.055
13	.259	.538	.870	1.350	1.771	2.160	2.650	3.012
14	.258	.537	.868	1.345	1.761	2.145	2.624	2.977
15	.258	.536	.866	1.341	1.753	2.131	2.602	2.947
16	.258	.535	.865	1.337	1.746	2.120	2.583	2.921
17	.257	.534	.863	1.333	1.740	2.110	2.567	2.898
18	.257	.534	.862	1.330	1.734	2.101	2.552	2.878
19	.257	.533	.861	1.328	1.729	2.093	2.539	2.861
20	.257	.533	.860	1.325	1.725	2.086	2.528	2.845
21	.257	.532	.859	1.323	1.721	2.080	2.518	2.831
22	.256	.532	.858	1.321	1.717	2.074	2.508	2.819
23	.256	.532	.858	1.319	1.714	2.069	2.500	2.807
24	.256	.531	.857	1.318	1.711	2.064	2.492	2.797
25	.256	.531	.856	1.316	1.708	2.060	2.485	2.787
26	.256	.531	.856	1.315	1.706	2.056	2.479	2.779
27	.256	.531	.855	1.314	1.703	2.052	2.473	2.771
28	.256	.530	.855	1.313	1.701	2.048	2.467	2.763
29	.256	.530	.854	1.311	1.699	2.045	2.462	2.756
30	.256	.530	.854	1.310	1.697	2.042	2.457	2.750
40	.255	.529	.851	1.303	1.684	2.021	2.423	2.704
60	.254	.527	.848	1.296	1.671	2.000	2.390	2.660
120	.254	.526	.845	1.289	1.658	1.980	2.358	2.617
∞	.253	.524	.842	1.282	1.645	1.960	2.326	2.576
df	$-t_{.40}$	$-t_{.30}$	$-t_{.20}$	$-t_{.10}$	$-t_{.05}$	$-t_{.025}$	$-t_{.01}$	$-t_{.005}$

When the table is read from the foot, the tabled values are to be prefixed with a negative sign. Interpolation should be performed using the reciprocals of the degrees of freedom.

Table A-1 is abridged from Table III of Fisher and Yates: "Statistical Tables for Biological Agricultural and Medical Research" published by Oliver and Boyd, Ltd., Edinburgh, and by permission of the authors and publishers.

TABLE A-2
PERCENTILES OF THE χ^2 DISTRIBUTION

df	Per Cent									
	.5	1	2.5	5	10	90	95	97.5	99	99.5
1	.000039	.00016	.00098	.0039	.0158	2.71	3.84	5.02	6.63	7.88
2	.0100	.0201	.0506	.1026	.2107	4.61	5.99	7.38	9.21	10.60
3	.0717	.115	.216	.352	.584	6.25	7.81	9.35	11.34	12.84
4	.207	.297	.484	.711	1.064	7.78	9.49	11.14	13.28	14.86
5	.412	.554	.831	1.15	1.61	9.24	11.07	12.83	15.09	16.75
6	.676	.872	1.24	1.64	2.20	10.64	12.59	14.45	16.81	18.55
7	.989	1.24	1.69	2.17	2.83	12.02	14.07	16.01	18.48	20.28
8	1.34	1.65	2.18	2.73	3.49	13.36	15.51	17.53	20.09	21.96
9	1.73	2.09	2.70	3.33	4.17	14.68	16.92	19.02	21.67	23.59
10	2.16	2.56	3.25	3.94	4.87	15.99	18.31	20.48	23.21	25.19
11	2.60	3.05	3.82	4.57	5.58	17.28	19.68	21.92	24.73	26.76
12	3.07	3.57	4.40	5.23	6.30	18.55	21.03	23.34	26.22	28.30
13	3.57	4.11	5.01	5.89	7.04	19.81	22.36	24.74	27.69	29.82
14	4.07	4.66	5.63	6.57	7.79	21.06	23.68	26.12	29.14	31.32
15	4.60	5.23	6.26	7.26	8.55	22.31	25.00	27.49	30.58	32.80
16	5.14	5.81	6.91	7.96	9.31	23.54	26.30	28.85	32.00	34.27
18	6.26	7.01	8.23	9.39	10.86	25.99	28.87	31.53	34.81	37.16
20	7.43	8.26	9.59	10.85	12.44	28.41	31.41	34.17	37.57	40.00
24	9.89	10.86	12.40	13.85	15.66	33.20	36.42	39.36	42.98	45.56
30	13.79	14.95	16.79	18.49	20.60	40.26	43.77	46.98	50.89	53.67
40	20.71	22.16	24.43	26.51	29.05	51.81	55.76	59.34	63.69	66.77
60	35.53	37.48	40.48	43.19	46.46	74.40	79.08	83.30	88.38	91.95
120	83.85	86.92	91.58	95.70	100.62	140.23	146.57	152.21	158.95	163.64

For large values of degrees of freedom the approximate formula

$$\chi_{\alpha}^{2} = n\left(1 - \frac{2}{9n} + z_{\alpha}\sqrt{\frac{2}{9n}}\right)^{3}$$

where z_α is the normal deviate and n is the number of degrees of freedom, may be used. For example $\chi_{.99}^2 = 60[1 - .00370 + 2.326(.06086)]^3 = 60(1.1379)^3 = 88.4$ for the 99th percentile for 60 degrees of freedom.

From *Introduction to Statistical Analysis* by Wilfrid J. Dixon and Frank J. Massey, Jr., Second Edition (1957) McGraw-Hill Book Co., Inc. Used by permission.

TABLE A-3
F DISTRIBUTION, UPPER 5% POINTS ($F_{.95}$)

Degrees of freedom for numerator (columns); Degrees of freedom for denominator (rows)

	1	2	3	4	5	6	7	8	9	10	12	15	20	24	30	40	60	120	∞
1	161	200	216	225	230	234	237	239	241	242	244	246	248	249	250	251	252	253	254
2	18.5	19.0	19.2	19.2	19.3	19.3	19.4	19.4	19.4	19.4	19.4	19.4	19.4	19.5	19.5	19.5	19.5	19.5	19.5
3	10.1	9.55	9.28	9.12	9.01	8.94	8.89	8.85	8.81	8.79	8.74	8.70	8.66	8.64	8.62	8.59	8.57	8.55	8.53
4	7.71	6.94	6.59	6.39	6.26	6.16	6.09	6.04	6.00	5.96	5.91	5.86	5.80	5.77	5.75	5.72	5.69	5.66	5.63
5	6.61	5.79	5.41	5.19	5.05	4.95	4.88	4.82	4.77	4.74	4.68	4.62	4.56	4.53	4.50	4.46	4.43	4.40	4.37
6	5.99	5.14	4.76	4.53	4.39	4.28	4.21	4.15	4.10	4.06	4.00	3.94	3.87	3.84	3.81	3.77	3.74	3.70	3.67
7	5.59	4.74	4.35	4.12	3.97	3.87	3.79	3.73	3.68	3.64	3.57	3.51	3.44	3.41	3.38	3.34	3.30	3.27	3.23
8	5.32	4.46	4.07	3.84	3.69	3.58	3.50	3.44	3.39	3.35	3.28	3.22	3.15	3.12	3.08	3.04	3.01	2.97	2.93
9	5.12	4.26	3.86	3.63	3.48	3.37	3.29	3.23	3.18	3.14	3.07	3.01	2.94	2.90	2.86	2.83	2.79	2.75	2.71
10	4.96	4.10	3.71	3.48	3.33	3.22	3.14	3.07	3.02	2.98	2.91	2.85	2.77	2.74	2.70	2.66	2.62	2.58	2.54
11	4.84	3.98	3.59	3.36	3.20	3.09	3.01	2.95	2.90	2.85	2.79	2.72	2.65	2.61	2.57	2.53	2.49	2.45	2.40
12	4.75	3.89	3.49	3.26	3.11	3.00	2.91	2.85	2.80	2.75	2.69	2.62	2.54	2.51	2.47	2.43	2.38	2.34	2.30
13	4.67	3.81	3.41	3.18	3.03	2.92	2.83	2.77	2.71	2.67	2.60	2.53	2.46	2.42	2.38	2.34	2.30	2.25	2.21
14	4.60	3.74	3.34	3.11	2.96	2.85	2.76	2.70	2.65	2.60	2.53	2.46	2.39	2.35	2.31	2.27	2.22	2.18	2.13
15	4.54	3.68	3.29	3.06	2.90	2.79	2.71	2.64	2.59	2.54	2.48	2.40	2.33	2.29	2.25	2.20	2.16	2.11	2.07
16	4.49	3.63	3.24	3.01	2.85	2.74	2.66	2.59	2.54	2.49	2.42	2.35	2.28	2.24	2.19	2.15	2.11	2.06	2.01
17	4.45	3.59	3.20	2.96	2.81	2.70	2.61	2.55	2.49	2.45	2.38	2.31	2.23	2.19	2.15	2.10	2.06	2.01	1.96
18	4.41	3.55	3.16	2.93	2.77	2.66	2.58	2.51	2.46	2.41	2.34	2.27	2.19	2.15	2.11	2.06	2.02	1.97	1.92
19	4.38	3.52	3.13	2.90	2.74	2.63	2.54	2.48	2.42	2.38	2.31	2.23	2.16	2.11	2.07	2.03	1.98	1.93	1.88
20	4.35	3.49	3.10	2.87	2.71	2.60	2.51	2.45	2.39	2.35	2.28	2.20	2.12	2.08	2.04	1.99	1.95	1.90	1.84
21	4.32	3.47	3.07	2.84	2.68	2.57	2.49	2.42	2.37	2.32	2.25	2.18	2.10	2.05	2.01	1.96	1.92	1.87	1.81
22	4.30	3.44	3.05	2.82	2.66	2.55	2.46	2.40	2.34	2.30	2.23	2.15	2.07	2.03	1.98	1.94	1.89	1.84	1.78
23	4.28	3.42	3.03	2.80	2.64	2.53	2.44	2.37	2.32	2.27	2.20	2.13	2.05	2.01	1.96	1.91	1.86	1.81	1.76
24	4.26	3.40	3.01	2.78	2.62	2.51	2.42	2.36	2.30	2.25	2.18	2.11	2.03	1.98	1.94	1.89	1.84	1.79	1.73
25	4.24	3.39	2.99	2.76	2.60	2.49	2.40	2.34	2.28	2.24	2.16	2.09	2.01	1.96	1.92	1.87	1.82	1.77	1.71
30	4.17	3.32	2.92	2.69	2.53	2.42	2.33	2.27	2.21	2.16	2.09	2.01	1.93	1.89	1.84	1.79	1.74	1.68	1.62
40	4.08	3.23	2.84	2.61	2.45	2.34	2.25	2.18	2.12	2.08	2.00	1.92	1.84	1.79	1.74	1.69	1.64	1.58	1.51
60	4.00	3.15	2.76	2.53	2.37	2.25	2.17	2.10	2.04	1.99	1.92	1.84	1.75	1.70	1.65	1.59	1.53	1.47	1.39
120	3.92	3.07	2.68	2.45	2.29	2.18	2.09	2.02	1.96	1.91	1.83	1.75	1.66	1.61	1.55	1.50	1.43	1.35	1.25
∞	3.84	3.00	2.60	2.37	2.21	2.10	2.01	1.94	1.88	1.83	1.75	1.67	1.57	1.52	1.46	1.39	1.32	1.22	1.00

Interpolation should be performed using reciprocals of the degrees of freedom.

TABLE A-4
F DISTRIBUTION, UPPER 1% POINTS ($F_{.99}$)

Degrees of freedom for numerator (columns); Degrees of freedom for denominator (rows)

	1	2	3	4	5	6	7	8	9	10	12	15	20	24	30	40	60	120	∞
1	4,052	5,000	5,403	5,625	5,764	5,859	5,928	5,982	6,023	6,056	6,106	6,157	6,209	6,235	6,261	6,287	6,313	6,339	6,366
2	98.5	99.0	99.2	99.2	99.3	99.3	99.4	99.4	99.4	99.4	99.4	99.4	99.4	99.5	99.5	99.5	99.5	99.5	99.5
3	34.1	30.8	29.5	28.7	28.2	27.9	27.7	27.5	27.3	27.2	27.1	26.9	26.7	26.6	26.5	26.4	26.3	26.2	26.1
4	21.2	18.0	16.7	16.0	15.5	15.2	15.0	14.8	14.7	14.5	14.4	14.2	14.0	13.9	13.8	13.7	13.7	13.6	13.5
5	16.3	13.3	12.1	11.4	11.0	10.7	10.5	10.3	10.2	10.1	9.89	9.72	9.55	9.47	9.38	9.29	9.20	9.11	9.02
6	13.7	10.9	9.78	9.15	8.75	8.47	8.26	8.10	7.98	7.87	7.72	7.56	7.40	7.31	7.23	7.14	7.06	6.97	6.88
7	12.2	9.55	8.45	7.85	7.46	7.19	6.99	6.84	6.72	6.62	6.47	6.31	6.16	6.07	5.99	5.91	5.82	5.74	5.65
8	11.3	8.65	7.59	7.01	6.63	6.37	6.18	6.03	5.91	5.81	5.67	5.52	5.36	5.28	5.20	5.12	5.03	4.95	4.86
9	10.6	8.02	6.99	6.42	6.06	5.80	5.61	5.47	5.35	5.26	5.11	4.96	4.81	4.73	4.65	4.57	4.48	4.40	4.31
10	10.0	7.56	6.55	5.99	5.64	5.39	5.20	5.06	4.94	4.85	4.71	4.56	4.41	4.33	4.25	4.17	4.08	4.00	3.91
11	9.65	7.21	6.22	5.67	5.32	5.07	4.89	4.74	4.63	4.54	4.40	4.25	4.10	4.02	3.94	3.86	3.78	3.69	3.60
12	9.33	6.93	5.95	5.41	5.06	4.82	4.64	4.50	4.39	4.30	4.16	4.01	3.86	3.78	3.70	3.62	3.54	3.45	3.36
13	9.07	6.70	5.74	5.21	4.86	4.62	4.44	4.30	4.19	4.10	3.96	3.82	3.66	3.59	3.51	3.43	3.34	3.25	3.17
14	8.86	6.51	5.56	5.04	4.70	4.46	4.28	4.14	4.03	3.94	3.80	3.66	3.51	3.43	3.35	3.27	3.18	3.09	3.00
15	8.68	6.36	5.42	4.89	4.56	4.32	4.14	4.00	3.89	3.80	3.67	3.52	3.37	3.29	3.21	3.13	3.05	2.96	2.87
16	8.53	6.23	5.29	4.77	4.44	4.20	4.03	3.89	3.78	3.69	3.55	3.41	3.26	3.18	3.10	3.02	2.93	2.84	2.75
17	8.40	6.11	5.19	4.67	4.34	4.10	3.93	3.79	3.68	3.59	3.46	3.31	3.16	3.08	3.00	2.92	2.83	2.75	2.65
18	8.29	6.01	5.09	4.58	4.25	4.01	3.84	3.71	3.60	3.51	3.37	3.23	3.08	3.00	2.92	2.84	2.75	2.66	2.57
19	8.19	5.93	5.01	4.50	4.17	3.94	3.77	3.63	3.52	3.43	3.30	3.15	3.00	2.92	2.84	2.76	2.67	2.58	2.49
20	8.10	5.85	4.94	4.43	4.10	3.87	3.70	3.56	3.46	3.37	3.23	3.09	2.94	2.86	2.78	2.69	2.61	2.52	2.42
21	8.02	5.78	4.87	4.37	4.04	3.81	3.64	3.51	3.40	3.31	3.17	3.03	2.88	2.80	2.72	2.64	2.55	2.46	2.36
22	7.95	5.72	4.82	4.31	3.99	3.76	3.59	3.45	3.35	3.26	3.12	2.98	2.83	2.75	2.67	2.58	2.50	2.40	2.31
23	7.88	5.66	4.76	4.26	3.94	3.71	3.54	3.41	3.30	3.21	3.07	2.93	2.78	2.70	2.62	2.54	2.45	2.35	2.26
24	7.82	5.61	4.72	4.22	3.90	3.67	3.50	3.36	3.26	3.17	3.03	2.89	2.74	2.66	2.58	2.49	2.40	2.31	2.21
25	7.77	5.57	4.68	4.18	3.86	3.63	3.46	3.32	3.22	3.13	2.99	2.85	2.70	2.62	2.53	2.45	2.36	2.27	2.17
30	7.56	5.39	4.51	4.02	3.70	3.47	3.30	3.17	3.07	2.98	2.84	2.70	2.55	2.47	2.39	2.30	2.21	2.11	2.01
40	7.31	5.18	4.31	3.83	3.51	3.29	3.12	2.99	2.89	2.80	2.66	2.52	2.37	2.29	2.20	2.11	2.02	1.92	1.80
60	7.08	4.98	4.13	3.65	3.34	3.12	2.95	2.82	2.72	2.63	2.50	2.35	2.20	2.12	2.03	1.94	1.84	1.73	1.60
120	6.85	4.79	3.95	3.48	3.17	2.96	2.79	2.66	2.56	2.47	2.34	2.19	2.03	1.95	1.86	1.76	1.66	1.53	1.38
∞	6.63	4.61	3.78	3.32	3.02	2.80	2.64	2.51	2.41	2.32	2.18	2.04	1.88	1.79	1.70	1.59	1.47	1.32	1.00

Interpolation should be performed using reciprocals of the degrees of freedom.

Reproduced from *Biometrika*, Vol. 33 (1943), p. 73. With permission from the *Biometrika* Trustees.

TABLE A-5
PERCENTILES OF THE DISTRIBUTION OF $q = w/s$,
UPPER 5% POINTS ($q_{.95}$)

ν \ k	2	3	4	5	6	7	8	9	10
1	17.97	26.98	32.82	37.08	40.41	43.12	45.40	47.36	49.07
2	6.08	8.33	9.80	10.88	11.74	12.44	13.03	13.54	13.99
3	4.50	5.91	6.82	7.50	8.04	8.48	8.85	9.18	9.46
4	3.93	5.04	5.76	6.29	6.71	7.05	7.35	7.60	7.83
5	3.64	4.60	5.22	5.67	6.03	6.33	6.58	6.80	6.99
6	3.46	4.34	4.90	5.30	5.63	5.90	6.12	6.32	6.49
7	3.34	4.16	4.68	5.06	5.36	5.61	5.82	6.00	6.16
8	3.26	4.04	4.53	4.89	5.17	5.40	5.60	5.77	5.92
9	3.20	3.95	4.41	4.76	5.02	5.24	5.43	5.59	5.74
10	3.15	3.88	4.33	4.65	4.91	5.12	5.30	5.46	5.60
11	3.11	3.82	4.26	4.57	4.82	5.03	5.20	5.35	5.49
12	3.08	3.77	4.20	4.51	4.75	4.95	5.12	5.27	5.39
13	3.06	3.73	4.15	4.45	4.69	4.88	5.05	5.19	5.32
14	3.03	3.70	4.11	4.41	4.64	4.83	4.99	5.13	5.25
15	3.01	3.67	4.08	4.37	4.59	4.78	4.94	5.08	5.20
16	3.00	3.65	4.05	4.33	4.56	4.74	4.90	5.03	5.15
17	2.98	3.63	4.02	4.30	4.52	4.70	4.86	4.99	5.11
18	2.97	3.61	4.00	4.28	4.49	4.67	4.82	4.96	5.07
19	2.96	3.59	3.98	4.25	4.47	4.65	4.79	4.92	5.04
20	2.95	3.58	3.96	4.23	4.45	4.62	4.77	4.90	5.01
24	2.92	3.53	3.90	4.17	4.37	4.54	4.68	4.81	4.92
30	2.89	3.49	3.85	4.10	4.30	4.46	4.60	4.72	4.82
40	2.86	3.44	3.79	4.04	4.23	4.39	4.52	4.63	4.73
60	2.83	3.40	3.74	3.98	4.16	4.31	4.44	4.55	4.65
120	2.80	3.36	3.68	3.92	4.10	4.24	4.36	4.47	4.56
∞	2.77	3.31	3.63	3.86	4.03	4.17	4.29	4.39	4.47

TABLE A-5 (Continued)
PERCENTILES OF THE DISTRIBUTION OF $q = w/s$,
UPPER 5% POINTS ($q_{.95}$)

ν \ k	11	12	13	14	15	16	17	18	19	20
1	50.59	51.96	53.20	54.33	55.36	56.32	57.22	58.04	58.83	59.56
2	14.39	14.75	15.08	15.38	15.65	15.91	16.14	16.37	16.57	16.77
3	9.72	9.95	10.15	10.35	10.52	10.69	10.84	10.98	11.11	11.24
4	8.03	8.21	8.37	8.52	8.66	8.79	8.91	9.03	9.13	9.23
5	7.17	7.32	7.47	7.60	7.72	7.83	7.93	8.03	8.12	8.21
6	6.65	6.79	6.92	7.03	7.14	7.24	7.34	7.43	7.51	7.59
7	6.30	6.43	6.55	6.66	6.76	6.85	6.94	7.02	7.10	7.17
8	6.05	6.18	6.29	6.39	6.48	6.57	6.65	6.73	6.80	6.87
9	5.87	5.98	6.09	6.19	6.28	6.36	6.44	6.51	6.58	6.64
10	5.72	5.83	5.93	6.03	6.11	6.19	6.27	6.34	6.40	6.47
11	5.61	5.71	5.81	5.90	5.98	6.06	6.13	6.20	6.27	6.33
12	5.51	5.61	5.71	5.80	5.88	5.95	6.02	6.09	6.15	6.21
13	5.43	5.53	5.63	5.71	5.79	5.86	5.93	5.99	6.05	6.11
14	5.36	5.46	5.55	5.64	5.71	5.79	5.85	5.91	5.97	6.03
15	5.31	5.40	5.49	5.57	5.65	5.72	5.78	5.85	5.90	5.96
16	5.26	5.35	5.44	5.52	5.59	5.66	5.73	5.79	5.84	5.90
17	5.21	5.31	5.39	5.47	5.54	5.61	5.67	5.73	5.79	5.84
18	5.17	5.27	5.35	5.43	5.50	5.57	5.63	5.69	5.74	5.79
19	5.14	5.23	5.31	5.39	5.46	5.53	5.59	5.65	5.70	5.75
20	5.11	5.20	5.28	5.36	5.43	5.49	5.55	5.61	5.66	5.71
24	5.01	5.10	5.18	5.25	5.32	5.38	5.44	5.49	5.55	5.59
30	4.92	5.00	5.08	5.15	5.21	5.27	5.33	5.38	5.43	5.47
40	4.82	4.90	4.98	5.04	5.11	5.16	5.22	5.27	5.31	5.36
60	4.73	4.81	4.88	4.94	5.00	5.06	5.11	5.15	5.20	5.24
120	4.64	4.71	4.78	4.84	4.90	4.95	5.00	5.04	5.09	5.13
∞	4.55	4.62	4.68	4.74	4.80	4.85	4.89	4.93	4.97	5.01

w is the range of k observations, and ν is the number of degrees of freedom in the independent standard deviation s.

TABLE A-5 (Continued)
PERCENTILES OF THE DISTRIBUTION OF q = w/s, UPPER 1% POINTS ($q_{.99}$)

ν \ k	2	3	4	5	6	7	8	9	10
1	90.03	135.0	164.3	185.6	202.2	215.8	227.2	237.0	245.6
2	14.04	19.02	22.29	24.72	26.63	28.20	29.53	30.68	31.69
3	8.26	10.62	12.17	13.33	14.24	15.00	15.64	16.20	16.69
4	6.51	8.12	9.17	9.96	10.58	11.10	11.55	11.93	12.27
5	5.70	6.98	7.80	8.42	8.91	9.32	9.67	9.97	10.24
6	5.24	6.33	7.03	7.56	7.97	8.32	8.61	8.87	9.10
7	4.95	5.92	6.54	7.01	7.37	7.68	7.94	8.17	8.37
8	4.75	5.64	6.20	6.62	6.96	7.24	7.47	7.68	7.86
9	4.60	5.43	5.96	6.35	6.66	6.91	7.13	7.33	7.49
10	4.48	5.27	5.77	6.14	6.43	6.67	6.87	7.05	7.21
11	4.39	5.15	5.62	5.97	6.25	6.48	6.67	6.84	6.99
12	4.32	5.05	5.50	5.84	6.10	6.32	6.51	6.67	6.81
13	4.26	4.96	5.40	5.73	5.98	6.19	6.37	6.53	6.67
14	4.21	4.89	5.32	5.63	5.88	6.08	6.26	6.41	6.54
15	4.17	4.84	5.25	5.56	5.80	5.99	6.16	6.31	6.44
16	4.13	4.79	5.19	5.49	5.72	5.92	6.08	6.22	6.35
17	4.10	4.74	5.14	5.43	5.66	5.85	6.01	6.15	6.27
18	4.07	4.70	5.09	5.38	5.60	5.79	5.94	6.08	6.20
19	4.05	4.67	5.05	5.33	5.55	5.73	5.89	6.02	6.14
20	4.02	4.64	5.02	5.29	5.51	5.69	5.84	5.97	6.09
24	3.96	4.55	4.91	5.17	5.37	5.54	5.69	5.81	5.92
30	3.89	4.45	4.80	5.05	5.24	5.40	5.54	5.65	5.76
40	3.82	4.37	4.70	4.93	5.11	5.26	5.39	5.50	5.60
60	3.76	4.28	4.59	4.82	4.99	5.13	5.25	5.36	5.45
120	3.70	4.20	4.50	4.71	4.87	5.01	5.12	5.21	5.30
∞	3.64	4.12	4.40	4.60	4.76	4.88	4.99	5.08	5.16

TABLE A-5 (Continued)
PERCENTILES OF THE DISTRIBUTION OF $q = w/s$, UPPER 1% POINTS ($q_{.99}$)

ν \ k	11	12	13	14	15	16	17	18	19	20
1	253.2	260.0	266.2	271.8	277.0	281.8	286.3	290.4	294.3	298.0
2	32.59	33.40	34.13	34.81	35.43	36.00	36.53	37.03	37.50	37.95
3	17.13	17.53	17.89	18.22	18.52	18.81	19.07	19.32	19.55	19.77
4	12.57	12.84	13.09	13.32	13.53	13.73	13.91	14.08	14.24	14.40
5	10.48	10.70	10.89	11.08	11.24	11.40	11.55	11.68	11.81	11.93
6	9.30	9.48	9.65	9.81	9.95	10.08	10.21	10.32	10.43	10.54
7	8.55	8.71	8.86	9.00	9.12	9.24	9.35	9.46	9.55	9.65
8	8.03	8.18	8.31	8.44	8.55	8.66	8.76	8.85	8.94	9.03
9	7.65	7.78	7.91	8.03	8.13	8.23	8.33	8.41	8.49	8.57
10	7.36	7.49	7.60	7.71	7.81	7.91	7.99	8.08	8.15	8.23
11	7.13	7.25	7.36	7.46	7.56	7.65	7.73	7.81	7.88	7.95
12	6.94	7.06	7.17	7.26	7.36	7.44	7.52	7.59	7.66	7.73
13	6.79	6.90	7.01	7.10	7.19	7.27	7.35	7.42	7.48	7.55
14	6.66	6.77	6.87	6.96	7.05	7.13	7.20	7.27	7.33	7.39
15	6.55	6.66	6.76	6.84	6.93	7.00	7.07	7.14	7.20	7.26
16	6.46	6.56	6.66	6.74	6.82	6.90	6.97	7.03	7.09	7.15
17	6.38	6.48	6.57	6.66	6.73	6.81	6.87	6.94	7.00	7.05
18	6.31	6.41	6.50	6.58	6.65	6.73	6.79	6.85	6.91	6.97
19	6.25	6.34	6.43	6.51	6.58	6.65	6.72	6.78	6.84	6.89
20	6.19	6.28	6.37	6.45	6.52	6.59	6.65	6.71	6.77	6.82
24	6.02	6.11	6.19	6.26	6.33	6.39	6.45	6.51	6.56	6.61
30	5.85	5.93	6.01	6.08	6.14	6.20	6.26	6.31	6.36	6.41
40	5.69	5.76	5.83	5.90	5.96	6.02	6.07	6.12	6.16	6.21
60	5.53	5.60	5.67	5.73	5.78	5.84	5.89	5.93	5.97	6.01
120	5.37	5.44	5.50	5.56	5.61	5.66	5.71	5.75	5.79	5.83
∞	5.23	5.29	5.35	5.40	5.45	5.49	5.54	5.57	5.61	5.65

TABLE A-6
FRACTION OF VARIANCE REMOVED BY CORRELATION (r^2)

	One independent variable		Two independent variables	
Significance level: df	0.95	0.99	0.95	0.99
1	0.994	1.000	0.998	1.000
2	0.902	0.980	0.951	0.990
3	0.771	0.920	0.865	0.953
4	0.658	0.841	0.776	0.901
5	0.569	0.764	0.699	0.841
6	0.500	0.696	0.632	0.785
8	0.399	0.585	0.527	0.684
10	0.332	0.501	0.450	0.602
12	0.283	0.437	0.393	0.536
15	0.232	0.367	0.329	0.458
20	0.179	0.288	0.259	0.370
25	0.145	0.237	0.213	0.308

After Volk, Chemical Engineering (March, 1956)

TABLE A-7
RANK CORRELATION COEFFICIENTS

Probability level: n	0.05	0.01
4	none	none
5	1.0	none
6	0.886	1.0
7	0.750	0.893
8	0.714	0.853
9	0.632	0.765
10	0.602	0.735
12	0.553	0.684
14	0.514	0.641
16	0.482	0.606
18	0.456	0.575
20	0.433	0.549

After Volk, Chemical Engineering (March, 1956)

APPENDIX B

Matrix methods

Nomenclature and General Application of Matrices

Though the nomenclature and application may seem rather complex at first, matrix procedures really are highly formalized and easy to follow.

An example of a matrix is

$$A = \begin{bmatrix} a_{11} & a_{12} \ldots a_{1n} \\ a_{21} & a_{22} \ldots a_{2n} \\ a_{m1} & a_{m2} \ldots a_{mn} \end{bmatrix}$$

The array has m rows and n columns, and would be called a matrix of $m \times n$ order.

For example, the two equations

$$\begin{aligned} a_{11}x + a_{12}y &= k_1 \\ a_{21}x + a_{21}y &= k_2 \end{aligned}$$

are represented by the matrix

$$\begin{bmatrix} a_{11} & a_{12} & k_1 \\ a_{21} & a_{22} & k_2 \end{bmatrix}$$

If there are n equations in n variables their matrix has n rows and $n + 1$ columns. The square array formed by a table of the n columns and n rows such as

$$\begin{bmatrix} a_{11} & a_{12} \\ a_{21} & a_{22} \end{bmatrix}$$

gives rise to a number called a *determinant* of the nth order and expressed analogously by use of vertical bars, thus:

$$\begin{vmatrix} a_{11} & a_{12} \\ a_{21} & a_{22} \end{vmatrix}$$

In this case, where $n = 2$, the determinant is merely another way of writing the difference of the diagonal products. This is

$$\begin{vmatrix} a_{11} & a_{12} \\ a_{21} & a_{22} \end{vmatrix} = a_{11}a_{22} - a_{12}a_{21}$$

The three determinants that form the elements of the example matrix are

$$D = \begin{vmatrix} a_{11} & a_{12} \\ a_{21} & a_{22} \end{vmatrix}, \quad D_x = \begin{vmatrix} k_1 & a_{12} \\ k_2 & a_{22} \end{vmatrix}, \quad D_y = \begin{vmatrix} a_{11} & k_1 \\ a_{21} & k_2 \end{vmatrix}$$

and

$$x = D_x/D; \; y = D_y/D$$

By orderly development of determinants the procedure of solving n simultaneous equations is reduced to a routine.

Matrices can be added, subtracted, multiplied, multiplied by a constant, transposed, and inverted.

Addition of two matrices is accomplished by adding corresponding elements; consequently addition is possible only for two matrices with the same number of rows and columns. For example,

$$\begin{bmatrix} 1 & 5 & 1 \\ 2 & 0 & 4 \\ 3 & 1 & 2 \end{bmatrix} + \begin{bmatrix} 4 & 2 & 0 \\ 1 & 3 & 1 \\ 2 & 0 & 2 \end{bmatrix} = \begin{bmatrix} 5 & 7 & 1 \\ 3 & 3 & 5 \\ 5 & 1 & 4 \end{bmatrix}$$

Subtraction is defined in a similar fashion.

Multiplication of a matrix by any constant means multiplying every element in the matrix by the constant, e.g.,

$$4 \begin{bmatrix} 1 & 2 & 1 \\ 2 & 5 & 3 \end{bmatrix} = \begin{bmatrix} 4 & 8 & 4 \\ 8 & 20 & 12 \end{bmatrix}$$

Given two matrices, the first with m_1 rows and n_1 columns and the second with m_2 rows and n_2 columns, their product is defined only if $n_1 = m_2$, i.e., the number of columns of the first equals the number of rows of the second. For two such matrices the elements of the product matrix are found by summing the element-by-element products of a row of the first matrix by a column of the second. For example.

$$A \times B = C$$

$$\begin{bmatrix} 4 & 1 & 3 \\ 2 & 2 & 1 \end{bmatrix} \times \begin{bmatrix} 1 & 5 \\ 2 & 5 \\ 1 & 1 \end{bmatrix} = \begin{bmatrix} 9 & 28 \\ 7 & 21 \end{bmatrix}$$

$$
\begin{aligned}
c_{11} &= a_{11}b_{11} + a_{12}b_{21} + a_{13}b_{31} = (4)(1) + (1)(2) + (3)(1) = 9 \\
c_{12} &= a_{11}b_{12} + a_{12}b_{22} + a_{13}b_{32} = (4)(5) + (1)(5) + (3)(1) = 28 \\
c_{21} &= a_{21}b_{11} + a_{22}b_{21} + a_{23}b_{31} = (2)(1) + (2)(2) + (1)(1) = 7 \\
c_{22} &= a_{21}b_{12} + a_{22}b_{22} + a_{23}b_{32} = (2)(5) + (2)(5) + (1)(1) = 21
\end{aligned}
$$

In matrix multiplication it is not necessarily true that AB = BA; order of multiplication usually makes a difference.

The transpose of a matrix, denoted by A′, is an $m \times n$ matrix whose rows are the columns of A and whose columns are the rows of A. If

$$A = \begin{bmatrix} 2 & 4 & 2 \\ 1 & 2 & 3 \end{bmatrix}$$

$$A' = \begin{bmatrix} 2 & 1 \\ 4 & 2 \\ 2 & 3 \end{bmatrix}$$

The inverse of a matrix is defined for square matrices only and is another matrix, A^{-1}, such that $A^{-1}A = I$ where the matrix I consists of ones down the main diagonal and zeros elsewhere, e.g., for a 4×4 matrix

$$I = \begin{bmatrix} 1 & 0 & 0 & 0 \\ 0 & 1 & 0 & 0 \\ 0 & 0 & 1 & 0 \\ 0 & 0 & 0 & 1 \end{bmatrix}$$

The inverse exists only if the determinant of the matrix is nonzero.

Division of one matrix by another is an inverse operation of multiplication. Let

$$A \times B = G$$

Then multiplying the equation on the left by A^{-1}

$$A^{-1} \times A \times B = A^{-1}G$$

but $A^{-1}A = I$, the identity matrix—which plays the same role in matrix algebra that 1 plays in ordinary algebra—so

$$B = A^{-1}G$$

This means that the operation of dividing matrix G by matrix A to get matrix B actually consists of multiplying matrix G by the inverse matrix of A.

As mentioned before, the elements of a square matrix define the determinant of a matrix. It may be expressed as a polynomial by expanding the determinant by minors according to Cramer's rule. A minor of an element of a determinant is the determinant of one less order found by striking out the row and column containing the particular element. The cofactor A_{ij} of the element a_{ij} is obtained by multiplying $(-1)^{i+j}$ by the minor of a_{ij}.

Cramer's rule permits evaluation of a determinant by multiplying the elements of any row or column by the corresponding cofactors and summing these products.

The element a^{ij} of the inverse A^{-1} of the matrix A is obtained by dividing the cofactor A_{ij} by the determinant $|A|$ of the matrix, that is,

$$a^{ij} = A_{ji}/|A|$$

For example, in order to find the inverse A^{-1} of the matrix

$$A = \begin{bmatrix} 0 & 3 & 2 \\ 1 & 2 & 4 \\ 3 & 0 & 2 \end{bmatrix}$$

it would be necessary to evaluate the determinant

$$|A| = \begin{vmatrix} 0 & 3 & 2 \\ 1 & 2 & 4 \\ 3 & 0 & 2 \end{vmatrix} = 0 \times \begin{vmatrix} 2 & 4 \\ 0 & 2 \end{vmatrix} - 3 \times \begin{vmatrix} 1 & 4 \\ 3 & 2 \end{vmatrix} + 2 \times \begin{vmatrix} 1 & 2 \\ 3 & 0 \end{vmatrix}$$

$$= 0 - 3(2 - 12) + 2(0 - 6) = 18$$

The cofactors are

$$A_{11} = (-1)^{1+1} = \begin{vmatrix} 2 & 4 \\ 0 & 2 \end{vmatrix} = 4$$

$$A_{12} = (-1)^{1+2} = \begin{vmatrix} 1 & 4 \\ 3 & 2 \end{vmatrix} = 10, \text{ etc.}$$

The complete matrix of cofactors is found to be

$$A_{ij} = \begin{bmatrix} 4 & 10 & -6 \\ -6 & -6 & 9 \\ 8 & 2 & -3 \end{bmatrix}$$

Upon interchanging rows and columns and dividing each element by A, the inverse matrix is found to be

$$A^{-1} = \begin{bmatrix} 4/18 & -6/18 & 8/18 \\ 10/18 & -6/18 & 2/18 \\ -6/18 & 9/18 & -3/18 \end{bmatrix}$$

It may be verified numerically that

$$A \times A^{-1} = A^{-1} \times A = I = \begin{bmatrix} 1 & 0 & 0 \\ 0 & 1 & 0 \\ 0 & 0 & 1 \end{bmatrix}$$

For example, the first element of the I matrix is

$$(0)(4/18) + (3)(10/18) + (2)(-6/18) = 1, \text{ etc.}$$

Least-squares regression—The matrix of normal equations

The least-squares principle is to minimize the sum of the squared deviations between the observed y values and those given by the regression equation. This is accomplished by taking the partial derivatives of

$$SSD = \sum_{i=1}^{N}(y_i - b_0x_{0i} - b_1x_{1i} - \cdots - b_px_{pi})^2$$

with respect to b_0, b_1, b_2, etc. and equating each of these derivatives to zero. The resulting system of *normal equations* has the form

$$\begin{array}{l}
b_0\Sigma x_0x_0 + b_1\Sigma x_0x_1 + b_2\Sigma x_0x_2 + \cdots + b_p\Sigma x_0x_p = \Sigma x_0y \\
b_0\Sigma x_1x_0 + b_1\Sigma x_1x_1 + b_2\Sigma x_1x_2 + \cdots + b_p\Sigma x_1x_p = \Sigma x_1y \\
b_0\Sigma x_2x_0 + b_1\Sigma x_2x_1 + b_2\Sigma x_2x_2 + \cdots + b_p\Sigma x_2x_p = \Sigma x_2y \\
\vdots \\
b_0\Sigma x_px_0 + b_1\Sigma x_px_1 + b_2\Sigma x_px_2 + \cdots + b_p\Sigma x_px_p = \Sigma x_py
\end{array}$$

In these equations x_0 is a dummy variable that is always equal to one.

It can be seen that the normal equations are expressible in matrix form as

$$A \times B = G$$

where

$$A = \begin{bmatrix}
\Sigma x_0^2 & \Sigma x_0x_1 & \Sigma x_0x_2 & \cdots & \Sigma x_0x_p \\
\Sigma x_1x_0 & \Sigma x_1^2 & \Sigma x_1x_2 & \cdots & \Sigma x_1x_p \\
\Sigma x_2x_0 & \Sigma x_2x_1 & \Sigma x_2^2 & \cdots & \Sigma x_2x_p \\
\cdots & \cdots & \cdots & \cdots & \cdots \\
\Sigma x_px_0 & \Sigma x_px_1 & \Sigma x_px_2 & \cdots & \Sigma x_p^2
\end{bmatrix}$$

$$B = \begin{bmatrix} b_0 \\ b_1 \\ b_2 \\ \cdot \\ \cdot \\ \cdot \\ b_p \end{bmatrix} \quad \text{and } G = \begin{bmatrix} \Sigma y \\ \Sigma x_1y \\ \Sigma x_2y \\ \cdot \\ \cdot \\ \cdot \\ \Sigma x_py \end{bmatrix}$$

The matrix A is a square symmetric matrix, while the matrices B and G are single-column matrices. Note that matrix A is symmetric around its principal diagonal and the elements of the principal diagonal are of the form Σx_i^2. The element in the ith row and jth column is Σx_ix_j, where the summation is over the N observations in these two variables. Clearly $\Sigma x_ix_j = \Sigma x_jx_i$, that is, the matrix is symmetric.

Mathematicians have shown that the solution for the matrix of regression coefficients is $B = A^{-1}G$.

Designating

$$A^{-1} = C = \begin{bmatrix} c_{00} & c_{01} & \cdots & c_{0r} \\ c_{10} & c_{11} & \cdots & c_{1r} \\ \cdots & \cdots & \cdots & \cdots \\ c_{r0} & c_{r1} & \cdots & c_{rr} \end{bmatrix}$$

then

$$\begin{bmatrix} b_0 \\ b_1 \\ \cdot \\ b_k \\ \cdot \\ b_y \end{bmatrix} = \begin{bmatrix} c_{00}c_{01} & \cdots & c_{0p} \\ c_{10}c_{11} & \cdots & c_{1p} \\ \cdots & \cdots & \cdots \\ c_{k0}c_{k1} & \cdots & c_{kp} \\ \cdots & \cdots & \cdots \\ c_{p0}c_{p1} & \cdots & c_{pp} \end{bmatrix} \times \begin{bmatrix} \Sigma y \\ \Sigma x_1 y \\ \cdot \\ \Sigma x_k y \\ \cdot \\ \Sigma x_p y \end{bmatrix}$$

Hence, since two matrices are equal if, (and only if) corresponding elements are equal,

$$b_k = c_{k0}\Sigma y + c_{k1}\Sigma x_1 y + \cdots + c_{kp}\Sigma x_p y$$

Instead of inverting matrix A directly, as illustrated earlier, in order to obtain the individual elements c_{ij} of the C-matrix, it is more convenient to proceed indirectly. Since $A \times A^{-1} = I$ and $A \times C = I$,

$$\begin{bmatrix} \Sigma x_0^2 & \Sigma x_0 x_1 & \cdots & \Sigma x_0 x_p \\ \Sigma x_1 x_0 & \Sigma x_1^2 & \cdots & \Sigma x_1 x_p \\ \cdots & \cdots & \cdots & \cdots \\ \Sigma x_p x_0 & \Sigma x_p x_1 & \cdots & \Sigma x_p^2 \end{bmatrix} \times \begin{bmatrix} c_{00}c_{01} & \cdots & c_{0p} \\ c_{10}c_{11} & \cdots & c_{1p} \\ \cdots & \cdots & \cdots \\ c_{p0}c_{p1} & \cdots & c_{pp} \end{bmatrix} = \begin{bmatrix} 1 & 0 & \cdots & 0 \\ 0 & 1 & \cdots & 0 \\ \cdots & \cdots & \cdots & \cdots \\ 0 & 0 & \cdots & 1 \end{bmatrix}$$

and, since two matrices are equal if, and only if, the corresponding elements are equal, it is immediately possible to write p sets of equations.

For example, the first set may be obtained by multiplying all the rows of the A matrix by the first column of the C matrix and setting the sums of these products in turn equal to the elements of the first column of the identity matrix, thus obtaining

$$\begin{aligned} c_{00}\Sigma x_0^2 + c_{10}\Sigma x_0 x_1 + \cdots + c_{p0}\Sigma x_0 x_p &= 1 \\ c_{00}\Sigma x_1 x_0 + c_{10}\Sigma x_1^2 + \cdots + c_{p0}\Sigma x_1 x_p &= 0 \\ \cdots\cdots\cdots\cdots\cdots\cdots\cdots\cdots & \\ c_{00}\Sigma x_p x_0 + c_{10}\Sigma x_p x_1 + \cdots + c_{p0}\Sigma x_p^2 &= 0 \end{aligned}$$

In a similar manner the other p-1 sets of equations can be obtained. The solution of these sets of equations results in the elements c_{ij} of the inverse matrix $C = A^{-1}$.

The Abbreviated Doolittle method

Where a large number of equations is involved, matrix inversion would be a laborious process, but for several short-cut methods that have been devised. Probably the most well known is the *Abbreviated Doolittle technique.* Actually, to get the elements of B (the regression coefficient matrix) it is necessary to carry through only the *forward portion* of the Abbreviated Doolittle method.

Table B-1 shows symbolically the steps involved in the forward portion of the Abbreviated Doolittle method for a system of five normal equations. Lines (1) to (5) contain the elements a_{ij} of the A-matrix. The lower left-hand corner of the matrix is omitted because of its symmetry with the upper right-hand portion. The elements g_i are elements of the G-matrix. The j column is used to check the computations and will be discussed later.

TABLE B-1
TABLE FOR ABBREVIATED DOOLITTLE METHOD

Line							
(1)	a_{11}	a_{12}	a_{13}	a_{14}	a_{15}	g_1	j_1
(2)		a_{22}	a_{23}	a_{24}	a_{25}	g_2	j_2
(3)			a_{33}	a_{34}	a_{35}	g_3	j_3
(4)				a_{44}	a_{45}	g_4	j_4
(5)					a_{55}	g_5	j_5
(6)	A_{11}	A_{12}	A_{13}	A_{14}	A_{15}	A_{1y}	j_6
(7)		B_{12}	B_{13}	B_{14}	B_{15}	B_{1y}	j_7
(8)		A_{22}	A_{23}	A_{24}	A_{25}	A_{2y}	j_8
(9)			B_{23}	B_{24}	B_{25}	B_{2y}	j_9
(10)			A_{33}	A_{34}	A_{35}	A_{3y}	j_{10}
(11)				B_{34}	B_{35}	B_{3y}	j_{11}
(12)				A_{44}	A_{45}	A_{4y}	j_{12}
(13)					B_{45}	B_{4y}	j_{13}
(14)					A_{55}	A_{5y}	j_{14}
(15)						B_{5y}	j_{15}

The solution starts by the recopying of line (1) as line (6), that is $A_{11} = a_{11}$, $A_{12} = a_{12}, \ldots A_{1y} = g_{1y}$. Line (7) is obtained from line (6) by dividing each number in line (6) by the leading element, A_{11}, that is

$$B_{12} = A_{12}/A_{11},\ B_{13} = A_{13}/A_{11},\ \text{etc.}$$

Line (8) is obtained by

$$\begin{aligned} A_{22} &= a_{22} - A_{12}B_{12} \\ A_{23} &= a_{23} - A_{12}B_{13} \\ A_{24} &= a_{24} - A_{12}B_{14} \\ A_{25} &= a_{25} - A_{12}B_{15} \\ A_{2y} &= g_2 \;\; - A_{12}B_{1y} \end{aligned}$$

Line (9) is obtained by

$$B_{23} = A_{23}/A_{22},\ B_{24} = A_{24}/A_{22}, \text{ etc.}$$

Line (10) is obtained by

$$\begin{aligned} A_{33} &= a_{33} - A_{13}B_{13} - A_{23}B_{23} \\ A_{34} &= a_{34} - A_{13}B_{14} - A_{23}B_{24} \\ A_{35} &= a_{35} - A_{13}B_{15} - A_{23}B_{25} \\ A_{3y} &= g_3 - A_{13}B_{1y} - A_{23}B_{2y} \end{aligned}$$

Line (11) is obtained by

$$B_{34} = A_{34}/A_{33},\ B_{35} = A_{35}/A_{33}, \text{ etc.}$$

Line (12) is obtained by

$$\begin{aligned} A_{44} &= a_{44} - A_{14}B_{14} - A_{24}B_{24} - A_{34}B_{34} \\ A_{45} &= a_{45} - A_{14}B_{15} - A_{24}B_{25} - A_{34}B_{35} \\ A_{4y} &= g_4 - A_{14}B_{1y} - A_{24}B_{2y} - A_{34}B_{3y} \end{aligned}$$

Line (13) is obtained by

$$B_{45} = A_{45}/A_{44}, \text{ etc.}$$

Line (14) is obtained by

$$\begin{aligned} A_{55} &= a_{55} - A_{15}B_{15} - A_{25}B_{25} - A_{35}B_{35} - A_{45}B_{45} \\ A_{5y} &= g_5 - A_{15}B_{1y} - A_{25}B_{2y} - A_{35}B_{3y} - A_{45}B_{4y} \end{aligned}$$

Line (15) is obtained by

$$B_{5y} = A_{5y}/A_{55}$$

When the point is reached where only a single B appears (here B_{5y}) the forward phase of the Doolittle is complete. The regression coefficients are now immediately available from

$$\begin{aligned} b_5 &= B_{5y} \\ b_4 &= B_{4y} - b_5B_{45} \\ b_3 &= B_{3y} - b_4B_{34} - b_5B_{35} \\ b_2 &= B_{2y} - b_3B_{23} - b_4B_{24} - b_5B_{25} \\ b_1 &= B_{1y} - b_2B_{12} - b_3B_{13} - b_4B_{14} - b_5B_{15} \end{aligned}$$

The j values are carried along as an arithmetic check on the rest of the work. The first six values are

$$\begin{aligned} j_1 &= a_{11} + a_{12} + a_{13} + a_{14} + a_{15} + g_1 \\ j_2 &= a_{12} + a_{22} + a_{23} + a_{24} + a_{25} + g_2 \\ j_3 &= a_{13} + a_{23} + a_{33} + a_{34} + a_{35} + g_3 \\ j_4 &= a_{14} + a_{24} + a_{34} + a_{44} + a_{45} + g_4 \\ j_5 &= a_{15} + a_{25} + a_{35} + a_{45} + a_{55} + g_5 \\ j_6 &= j_1 \end{aligned}$$

Subsequent j values are generated by operations analogous to those producing the A and B values, that is

$$
\begin{aligned}
j_7 &= j_6/A_{11}\\
j_8 &= j_2 - A_{12}j_7\\
j_9 &= j_8/A_{22}\\
j_{10} &= j_3 - A_{13}j_7 - A_{23}j_9\\
j_{11} &= j_{10}/A_{33}\\
j_{12} &= j_4 - A_{14}j_7 - A_{24}j_9 - A_{34}j_{11}\\
j_{13} &= j_{12}/A_{44}\\
j_{14} &= j_5 - A_{15}j_7 - A_{25}j_9 - A_{35}j_{11} - A_{45}j_{13}\\
j_{15} &= j_{14}/A_{55}
\end{aligned}
$$

If the j's are calculated in this way they serve as a check on the A's and B's as follows:

$$
\begin{aligned}
1 + B_{12} + B_{13} + B_{14} + B_{15} + B_{1y} &= j_7\\
A_{22} + A_{23} + A_{24} + A_{25} + A_{2y} &= j_8\\
1 + B_{23} + B_{24} + B_{25} + B_{2y} &= j_9\\
A_{33} + A_{34} + A_{35} + A_{3y} &= j_{10}\\
1 + B_{34} + B_{35} + B_{3y} &= j_{11}\\
A_{44} + A_{45} + A_{4y} &= j_{12}\\
1 + B_{45} + B_{4y} &= j_{13}\\
A_{55} + A_{5y} &= j_{14}\\
1 + B_{5y} &= j_{15}
\end{aligned}
$$

This checking system exposes immediately the errors made at each stage of the forward solution. A final over-all check on the b's can be made by substituting them back in the normal equations.

The elements of the inverse of A, denoted by c_{ij}, may now be found from the relations given below. Starting with c_{55}, the element of the lower right corner, work back towards c_{11}, as follows:

$$
\begin{aligned}
c_{55} &= 1/A_{55}\\
c_{45} &= -B_{45}c_{55}\\
c_{35} &= -B_{34}c_{45} - B_{35}c_{55}\\
c_{25} &= -B_{23}c_{35} - B_{24}c_{45} - B_{25}c_{55}\\
c_{15} &= -B_{12}c_{25} - B_{13}c_{35} - B_{14}c_{45} - B_{15}c_{55}\\
c_{44} &= 1/A_{44} - B_{45}c_{45}\\
c_{34} &= -B_{34}c_{44} - B_{35}c_{45}\\
c_{24} &= -B_{23}c_{34} - B_{24}c_{44} - B_{25}c_{45}\\
c_{14} &= -B_{12}c_{24} - B_{13}c_{34} - B_{14}c_{44} - B_{15}c_{45}\\
c_{33} &= 1/A_{33} - B_{34}c_{34} - B_{35}c_{35}\\
c_{23} &= -B_{23}c_{33} - B_{24}c_{34} - B_{25}c_{35}\\
c_{13} &= -B_{12}c_{23} - B_{13}c_{33} - B_{14}c_{34} - B_{15}c_{35}\\
c_{22} &= 1/A_{22} - B_{23}c_{23} - B_{24}c_{24} - B_{25}c_{25}\\
c_{12} &= -B_{12}c_{22} - B_{13}c_{23} - B_{14}c_{24} - B_{15}c_{25}\\
c_{11} &= 1/A_{11} - B_{12}c_{12} - B_{13}c_{13} - B_{14}c_{14} - B_{15}c_{15}
\end{aligned}
$$

This completes the calculation of the inverse matrix. For a symmetric matrix the inverse will also be symmetric so that the lower left corner need not be calculated. A final check should be made to see if $C \times A = 1$.

Index

Abbreviated Doolittle Technique, 101, 118, 156
Analysis of variance, 59
  balanced design, 69, 71
  cross-classification, 74, 78, 81
  hierarchic classification (see one-way classification)
  in curvilinear regression, 108
  in multiple regression, 120
  in orthogorial system, 133
  in regression analysis, 103
  model I, 83
  model II, 83
  one-way classification, 64, 84
  two-way classification, 74
  unbalanced design, 69, 70, 72, 81
Average (see Mean)

Backward Doolittle solution, 110, 120
Bartlett's test, 63, 67
β-Factors, 126
Bias, 35
Bimodal distribution, 20, 22, 31

Central tendency, 4, 9
Chi-square, 68
Chi-square, related to F, 86
Chi-square test, 50
  cross-classification (see Contingency test)
  equal expectancy, 52
  unequal expectancies, 54
Coding, 24, 26, 133
Coefficient of variation, 12, 15
Cofactor, 154
Comparison of means, 84
Confidence level, 43
  (see also Probability)
Confidence limits, 42
  on comparison of means, 86
  on estimate of the mean, 44
  on estimate of σ, 44–46
  on estimate of slope, 97
  on estimate of $\hat{y}$, 97, 104, 120, 123
  on multiple regression coefficients, 120, 121
  on prediction of $\hat{y}$, 98, 104, 129
  on regression coefficient, 103, 109
Contingency table, 55
Contingency test, 55
Correction for the mean, 66
Correlation analysis, 88
Correlation coefficient, 89, 90, 94, 98, 105, 121, 149
Covariance, 99, 109, 121, 131
Cramer's rule, 153
Cumulative normal curve, 20
Cumulative probability plot, 31, 33
Curvilinear regression, 105–112

Degrees of freedom, 14
  in chi-square test, 50, 52, 53
  in contingency test, 56
  in estimate of correlation coefficient, 90
  in F-test, 60
  in multiple correlation, 127
  in rank correlation, 93
  in t-test, 46, 48, 49, 50
  in Tukey's test, 85
Dependent variable (see variable, dependent)

Design of experiments, 130–137
(see also Factorial design)
Determinant, 101, 151
Differences in means, analysis of, 59
(see also Analysis of variance)
Dispersion, 4, 12
Distribution, 15
bimodal, 20, 22, 31
non-normal, 20, 21
normal, 15, 16, 29
of chi-square, 50, 142
of F, 60, 143, 144
of q, 85, 145–148
of rank correlation coefficient, 92, 149
of r^2, 89, 127, 149
of t, 43, 141
skew, 20, 21, 33, 45

Empirical relation, 94
Enumeration statistics, 50
Error variance, 61, 67, 97
(See also Variance due to experimental error)
Experimental error in curvilinear regression, 108

Factorial design, 132
F, related to chi-square, 86
F, related to t, 86
F-test, 60, 63, 67, 73, 77, 79, 83, 84, 103, 108, 120, 123, 135
Forward Doolittle solution, 102, 103, 106, 120, 129, 133, 137, 157
Frequency curve, 4
Frequency distribution, 4, 10, 11, 12
Functional relation, 88

Gaussion distribution, 15

Histogram, 4, 12

Independent variable (see Variable, independent)
Interaction, 77, 82
Interfactor correlation, 125, 126, 129
Inverse matrix, 122

Least squares, 95
Linear regression, 88, 89, 93, 94, 100, 155

Matrix, 101, 106, 151–159
addition, 152
inversion, 109, 110, 120, 121, 153, 156, 159

Matrix (*Continued*)
multiplication, 152
subtraction, 152
transposition, 153
Mean, 4, 5, 9, 19, 20, 26, 81
geometric, 8, 9
grand, 6, 61
grouped, 6, 85
harmonic, 8, 9, 81, 82
weighted, 6, 7, 24, 26
Mean square (see Variance)
Measurement error, 10, 69
Median, 9, 19, 20, 29
Mode, 9, 18, 19, 20, 29
Multiple correlation, 124–130
coefficient, 125, 126, 127, 129
Multiple regression, 116–124

Negative correlation, 92, 122
Normal equations, 118, 128, 155
Normal distribution, 15–19
Null hypothesis, 47, 51, 60

Orthogonality, 131, 132

Partial correlation, 125
Partial regression coefficient, 128
Percentile, 30, 113
Polynomial regression
(see Regression, curvilinear)
Population, 39, 40, 44, 63, 64, 84
mean, 39, 40, 42
standard deviation, 39, 40, 43
Positive correlation, 92
Precision, 10, 69, 77, 81
Primary correlation, 124, 126
coefficients, 129
Probability, 19, 29, 30, 46, 47, 51, 60, 67, 85, 93, 104, 129
density distribution, 4, 12, 19
Proportional frequency, 4

Random sample, 3
Random variation, 10
Range, 12, 13
Rank correlation, 92, 149
Replication, 78, 134
Regression, curvilinear, 105–112, 130, 136
coefficients, 107
experimental error, 108
Regression, linear, 88, 89, 93, 94, 100, 155
coefficient, 96, 129
graphical, 112
residual error, 95
standard error of estimate, 97

Regression, multiple, 116–124
  coefficients, 119, 122
  orthogonal system, 132
Robust tests, 39
Root mean square, 13
  (see also Standard deviation)

Sample, 3, 34, 35
Sampling, 3
  biased, 35, 36
  random, 36
Scatter diagram, 90, 91, 94
Significance tests, 46–83
  in linear regression, 97, 103, 109
  in multiple correlation, 126, 127
  in multiple regression, 120, 121, 123, 133–135
Simple correlation, 124, 129
Skewness (see Distribution, skew)
Standard deviation, 4, 12, 13, 26, 28, 84
  graphical determination of, 19, 114
Standard error, 74
Standard error of the mean, 39
Statistical collectives, 3
Statistical inferences, 3
Statistical population, 3, 18
Statistical universe, 3
Sum of squares, 61, 62
  about the mean (see sum of squares within sample)
  between samples, 62
  correction for the mean, 62
  crude, 62, 66
Sum of squares (*Continued*)
  pooled, 80
  total, 62
  within samples, 61
Systematic error, 68

t, related to F, 86
t test, 46, 47, 84, 109, 124
  one-sided, 48
  two-sided, 48
Transformation, 8, 22, 30, 101, 105, 113, 128, 130
Tukey's test, 85

Variable, 4
  dependent, 88, 91
  independent, 88, 91, 109, 123, 131, 132
Variance, 14, 26
  about the regression, 97
  due to experimental error, 61
  (see also Error variance)
  grouped, 63
  interaction, 77
  of estimated $\hat{y}$, 97, 104, 122
  of mean, 61, 97
  of predicted $\hat{y}$, 123
  of regression coefficient, 103, 109, 121
  of sample means, 61
  of slope, 97
  pooled, 63, 68
  within groups, 68, 77, 84

Weighting factor, 6, 7, 24, 25, 26, 28
ξ-functions, 136, 137